Sentinels in the Sky

Sentinels in the Sky

*How Satellites Have Revolutionized
Our Understanding of Earth and Space*

James Lawrence Powell

OXFORD
UNIVERSITY PRESS

OXFORD
UNIVERSITY PRESS

Oxford University Press is a department of the University of Oxford.
It furthers the University's objective of excellence in research, scholarship,
and education by publishing worldwide. Oxford is a registered trade mark of
Oxford University Press in the UK and in certain other countries.

Published in the United States of America by Oxford University Press
198 Madison Avenue, New York, NY 10016, United States of America.

CIP data is on file at the Library of Congress.

ISBN 9780197842829

DOI: 10.1093/9780197842850.001.0001

Printed by Integrated Books International, United States of America

The manufacturer's authorized representative in the EU for product safety is
Oxford University Press España S.A. of Parque Empresarial San Fernando de Henares,
Avenida de Castilla, 2 – 28830 Madrid (www.oup.es/en or product.safety@oup.com).
OUP España S.A. also acts as importer into Spain of products made by the manufacturer.

To Kemar, Amelia, Sophie, and Lawrence

Earth is the cradle of humanity but one cannot live in the cradle forever.

—Konstantin Tsiolkovsky
From a letter written in 1911[1]

Contents

1
Introduction

Few inventions have shaped history as profoundly as maps. They have been instrumental since the earliest civilizations of Babylonia and Egypt, who used them not only for practical purposes but also to showcase their knowledge, power, and influence. During the Age of Exploration, spanning the latter half of the fifteenth century through the sixteenth, European explorers, armed with increasingly accurate and comprehensive maps, reshaped the global economy and geopolitical landscape. The most significant map in history may have been the one created by Flemish cartographer Gerardus Mercator (1512–1594), who employed a new method of projection that depicted sailing routes as straight lines of constant bearing, which mariners needed only to follow to arrive at their destinations. Still in use today, Mercator's projection distorts polar regions—but this was of little concern to sixteenth-century sailors, who were not headed there. Maps have become ever more accurate and complete until today, one can turn to Google Earth to view a detailed map of the ocean floor.

Whereas maps reshaped space, clocks revolutionized time. The Egyptians and Greeks used sundials and water clocks to measure time based on natural phenomena. During the medieval period, these gradually gave way to mechanical clocks, initially located in church towers and public buildings. Consequently, people began to organize their lives—when to sleep, eat, pray, and work—according to the clock. We still answer to it. The invention of the pendulum clock by Christiaan Huygens in 1656 had a similar impact as Mercator's map, as it transformed timekeeping. Both had the advantage of being relatively simple to use.

Mapmaking and time measurement converged in the quest to determine a ship's longitude: the number of degrees east or west of the zero meridian at Greenwich, England, essential for pinpointing one's location at sea. Latitude, the distance measured north or south of the equator, is relatively straightforward to measure: just observe the angle of the sun or another celestial body above the horizon at a specific time and look up the corresponding latitude in a table. Longitude, however, presents a greater challenge. It requires knowing the time difference of your location from that of Greenwich, where

Sentinels in the Sky. James Lawrence Powell, Oxford University Press. © James Lawrence Powell (2026).
DOI: 10.1093/9780197842850.003.0001

each hour of difference equates to 15 degrees of longitude. On land, the issue can be resolved by using a clock accurately maintained at Greenwich time. At sea, though, the task becomes exceedingly difficult. Ships are in constant motion and are exposed to varying temperatures and humidity, conditions that make it nearly impossible to maintain an accurate timepiece. Another breakthrough came with John Harrison's (1693–1776) invention in 1761 of a marine chronometer that used temperature-compensated balances and friction-reducing devices to maintain accuracy despite the motion of ocean waves.

How things have changed since the days of Mercator, Huygens, and Harrison. Today, roughly five billion people worldwide can pinpoint their location to within a few meters and can tell time to a fraction of a second merely by glancing at their smartphone or smartwatch. According to one estimate, the number of smartphone users will rise to an estimated 6.5 billion by the end of the 2020s, compared to an estimated global population of 8.5 billion. In the United States today, 82 percent of the population owns a smartphone. A strong case can be made that in just a few years, smartphones have proven more revolutionary than Mercator's map and the clocks of Huygens and Harrison. What has made this possible? In a word: satellites.

The fundamental advantage of satellites is that they enable the simultaneous observation of large swaths of Earth's atmosphere and surface. The benefits of occupying the "high ground" have been acknowledged since the era of hunter-gatherers and the earliest treatises on the art of war. Before the satellite age, scientists seeking a bird's-eye view of Earth—to study weather patterns, for example—used airplanes that could reach an altitude of 15 km and balloons that carried instruments up to 30 km. An aircraft at 15-km altitude can survey an area roughly 700 km on a side, considerably smaller than most weather patterns. Once limited to a day or two, forecasts now extend to a fortnight—thanks to satellites and supercomputers.

Inspired by the imaginative works of Jules Verne and other science fiction authors, during the early twentieth century pioneering scientists in Germany, the USSR, and the United States developed principles that would enable rockets to be sent into space. Starting in the 1920s, they constructed and tested primitive models. Although these early rockets could not reach the velocity necessary for orbit, they demonstrated the potential for more advanced designs to do so. In the 1930s, two significant developments would profoundly affect rocketry. First, the prospect of another world war, just two decades after the conclusion of the first, became increasingly probable. Second, both rocket engineers and the military recognized that rockets could be scaled up to transport and deliver explosives to targets hundreds of kilometers away. Germany's head start in rocketry led to production of the V-1 "doodlebug," whose buzz tormented wartime Londoners in the early years of World War II. The V-1 carried an 850-kg payload at speeds of approximately 650 km per hour. In 1944, the more advanced V-2 arrived, boasting supersonic speeds, a heavier explosive payload, and a greater range. The trajectory of the V-2 briefly carried it beyond the boundary of space, but lacking the speed for orbit, it fell silently, striking targets without warning.

After World War II, the victorious United States and USSR, recognizing the potential of the V-2 rocket, began adapting it to transport weapons across continents and oceans. Using seized German rockets as a foundation, both nations also leveraged the expertise of German scientists who had been captured or surrendered. By the mid-1950s, both countries had developed rockets that, although initially intended for military use, were capable of propelling objects to orbital velocity. The Soviets understood that being the first nation in space would further demonstrate that in the realm of science and technology, the USSR was the equal of democratic governments. On October 4, 1957, the world awoke to the beep-beep signal from the orbiting Sputnik satellite, a metal sphere approximately twice the size of a basketball. The United States rushed to catch up, but the Soviets were first into space.

Then, a remarkable thing happened. While not dismissing the military applications of rockets, the United States started to recognize the immense potential of space exploration, including satellite observations, for both practical and scientific purposes. This culminated most spectacularly in the Apollo program, which sent astronauts to the Moon to collect and return specimens whose analysis helped scientists discover the Moon's origin.[1] Alongside these extraordinary achievements, under the auspices of the newly established National Aeronautics and Space Administration (NASA), established in 1958, the United States initiated a much less publicized, yet highly fruitful, scientific satellite program. Other nations gradually developed their own satellite capabilities: today eleven countries can launch satellites, while dozens more have space agencies but depend on one of the eleven to loft their satellites.

In short, satellites have transformed science, from tracking hurricanes to mapping the universe's origin. They monitor climate change, predict natural disasters, and tally Emperor Penguins in Antarctica. Today we are experiencing a golden age of satellite observation, with new applications continually emerging. Arguably, no other research program has combined such a wide range of practical applications with so many groundbreaking scientific discoveries.

PART I
THE ROCKET'S RED GLARE

2

Brick Moon to Peenemünde

Why does the Moon neither drift away nor crash into Earth? How does it manage to just hang there, unsupported? Isaac Newton (1643–1727) wrestled with this question for two decades before formulating his law of universal gravitation. Lacking the means to test his theory directly, he devised a thought experiment: a cannon mounted atop a high mountain, fired horizontally to illustrate the effect of gravity on orbital motion. If only enough gunpowder were used to launch the cannonball at a low velocity, gravity would soon pull it back to Earth. The more gunpowder used, the faster and farther the cannon-ball would travel before gravity brought it down. Newton's equations showed that at a speed of about 8,000 m/s, the force of gravity pulling the cannonball down would just balance the speed, pushing it forward, leaving it in perpetual orbit. With even more speed, the cannonball could escape Earth's gravity and fly off into space. Newton's law of universal gravitation, together with energy conservation, imply an escape velocity from Earth of 11,200 m/s.

The Brick Moon

In the mid-nineteenth century, achieving such speeds could only be imagined, so the earliest depictions of artificial satellites came from science fiction. These stories inspired the scientists who, in the following century, would develop the principles of astronautics (the science and technology of human space travel and exploration), which the next generation after them would use to build rockets capable of launching satellites and sending spacecraft to the Moon and beyond. The first of these works was the 1869 novella, *The Brick Moon*, by the American author Edward Everett Hale (1822–1909).[1] He described a group of scientists who constructed a hollow, 200-ft-diameter brick sphere intended for Earth orbit, as illustrated in Figure 2.1. Hale, a prominent abolitionist, was already renowned for his 1863 book, *The Man Without a Country*, the story of a traitor who, after joining Aaron Burr's trea-sonous and failed rebellion, had to renounce his American citizenship and spend the rest of his life at sea, stateless. During the U.S. Civil War, Hale's book

Sentinels in the Sky. James Lawrence Powell, Oxford University Press. © James Lawrence Powell (2026).
DOI: 10.1093/9780197842850.003.0002

Figure 2.1 The Brick Moon.

was instrumental in inspiring Northern patriotism. *The Brick Moon* appeared after the war to wide acclaim from hisloyal readership.

Hale envisioned the Brick Moon not as a vehicle for pleasure trips into space but as a tool to assist observers on Earth in calculating longitude. The novel imagined Earth surrounded by a ring, like Saturn's, aligned with the Greenwich prime meridian. The ring's position in the sky would vary based on longitude: the farther west from Greenwich one observed it, the lower the ring would appear in the sky. As Hale explained, "At Greenwich [the ring] would be directly overhead. At New Orleans [longitude 90°W], which is a quarter of the way around the world from Greenwich, it would be just at his horizon." Hale proposed that an orbiting brick sphere traveling along the prime meridian could replace the imagined ring, enabling observers on land or at sea to compute longitude by measuring the height of the sphere above the horizon.

Two colossal, energy-storing, water-powered flywheels (a heavy rotating disk that stores rotational energy and resists changes in speed, helping to smooth out fluctuations in mechanical systems) were to launch the sphere. In Hale's story, an accident inadvertently propels the Brick Moon into orbit while visiting families are aboard. The incident proves fortuitous, however, as the "Lunarians" discover greater happiness in orbit than the people they left behind on Earth. To communicate with their Earth-bound families, who could see them aboard their spacecraft but could not converse with them, the inventive Brick Mooners employed shorter and longer jumps to correspond to the dots and dashes of Morse code.

The Brick Moon functioned more as social satire on Hale's fellow Bostonians than as a work of science fiction, akin to the role the imaginary island of Lilliput played in Jonathan Swift's *Gulliver's Travels*. The significance of Hale's book for us is that the Brick Moon was not merely a stunt but had a practical purpose: to serve as a navigational satellite, making it conceptually a forerunner of today's Global Positioning System.

From the Earth to the Moon

Born a decade apart, Edward Everett Hale and Jules Verne each turned their imaginations skyward, envisioning journeys into space long before such voyages were possible. In his 1865 novel *From the Earth to the Moon*, Verne conceived of a giant cannon, the *Columbiad*, that could shoot an object into space, much like Newton's imaginary cannonball.[2] In Verne's tale, Civil War artillerymen repurpose their cannons for space travel. Verne provides rough calculations of the specifications the cannon would have to meet, and he addresses the question of how the three astronauts aboard the spacecraft would survive the tremendous acceleration the explosion would cause. Foreshadowing another risk of the space age, shortly after launch an asteroid narrowly misses the vessel.

At the end of the book, using a common literary tactic, Verne tantalizes his readers by leaving the astronauts' fate hanging, to be continued in the sequel: *Around the Moon*. In that later volume, instead of landing on the Moon as planned, the craft enters orbit around it, providing the astronauts with a bird's-eye view of the bright crater Tycho. They discuss the possibility of life on the Moon but, seeing no evidence, reject the notion. Most astronomers at the time still believed that the Moon harbored life, but Verne did not share that belief. The crew successfully returns to Earth, landing in the sea, where mariners rescue the floating capsule and the astronauts to worldwide acclaim. This event was mirrored in 1961 when one of the first American astronauts, Alan Shepard, splashed down in the Atlantic after a 15-minute suborbital flight in Mercury Spacecraft 7. Shepard was celebrated with ticker-tape parades and received the National Aeronautics and Space Administration (NASA) Distinguished Service Medal from President John F. Kennedy.

Long after his death, Jules Verne remains one of the most popular fiction writers in history. In 2017, the United Nations Educational, Scientific and Cultural Organization (UNESCO) ranked him second on the list of most-translated authors, trailing Agatha Christie and surpassing Shakespeare.[3] Verne's writings influenced three pioneers of rocketry and spaceflight, whom we will encounter next: Konstantin Tsiolkovsky (1857–1935), Robert H. Goddard (1882–1945), and Hermann Oberth (1894–1989).

Space Pioneers

As illustrated by Newton's thought experiment and his equations, placing an object into orbit around Earth required it to achieve cosmic velocities of thousands of meters per second. By the second half of the nineteenth century, a few visionaries had begun to imagine that an ancient weapon of war—the rocket—might be scaled up enough to reach such velocities and place an artificial satellite in orbit.

The origins of rocketry trace back to early China, where, in the thirteenth century, crude gunpowder-filled tubes were developed for military purposes. In 1232 CE, five years after Genghis Khan's death, the Chinese fired rockets to repel Mongol invaders during the Siege of Kai-fung-fu. This event marked one of the earliest recorded uses of rockets in warfare. Known as "fire arrows," these rudimentary missiles were used primarily as incendiary devices against enemy troops and fortifications. Soon, rocket technology spread to the Middle East and eventually to Europe.

Military rocketry continued to advance, with various nations contributing their own innovations. Among the most influential figures was Sir William

Congreve (1772–1828), an English inventor, artillery officer, and politician. Drawing on earlier Indian designs, Congreve developed a rocket with a sheet-metal casing, propelled by a combustible mixture of sulfur, charcoal, and potassium nitrate—essentially refined gunpowder.

Although the Congreve rocket was not available during the American Revolutionary War, it entered British military service during the Napoleonic Wars in the early nineteenth century. By the War of 1812, Congreve rockets had been further improved by the addition of a long stabilizing stick, which increased both their range and accuracy, though by modern standards they remained unpredictable.

One of the most enduring images of these weapons comes from the Battle of Baltimore, fought on September 13–14, 1814. From a British ship in the harbor, American attorney and amateur poet Francis Scott Key observed the bombardment of Fort McHenry. He later recalled the "red glare" of the Congreve rockets arching through the night sky—a scene that would inspire the lines of what became *The Star-Spangled Banner*. As dawn broke, the tattered but defiant American flag still flew over the fort (see Figure 2.2).

Figure 2.2 The Rocket's Red Glare.

MHS Licensing

The first person to propose using rocket propulsion for spaceflight was a Russian bomb expert, Nikolai Kibalchich (1853–1881), better known to history for being executed for his involvement in the assassination of Tsar Alexander II on March 13, 1881.[4] Kibalchich's ideas about rocketry were so advanced for his time that they could not be implemented, yet they contributed to an intellectual environment that may have inspired Konstantin Tsiolkovsky (see Figure 2.3), a self-taught Russian who became a leading authority on the physical and mathematical principles of rocket propulsion.

Tsiolkovsky lived his entire life in a small Russian town, where he devoted himself to teaching science and studying the principles of rocketry. A childhood bout of scarlet fever had left him partially deaf, and as a result, he worked almost exclusively alone, interacting with other Russian scientists through correspondence. Tsiolkovsky was an intriguing blend of reserved schoolteacher and passionate evangelist for space exploration. As his inspiration, he cited the work of Jules Verne on the mathematics of space rocketry. Tsiolkovsky himself became a prolific writer of science fiction andin turn, inspired the following generation of Russian scientists. The successors of those scientists would become the first to launch an artificial satellite, the

Figure 2.3 Konstantin Eduardovich Tsiolkovsky ~ 1924.

Wikimedia Commons

first to loft a human into space, and the first to send a spacecraft to the Moon. Despite his reclusiveness, Tsiolkovsky's pioneering rocketry earned him an award as a Hero of the Soviet Union and an invitation to deliver a speech on May Day 1935 to dignitaries assembled in Red Square, including Josef Stalin. The speech was taped and broadcast across the USSR. "Today, Comrades," Tsiolkovsky predicted, "I firmly believe that what were my past dreams—interplanetary travel—based solely on theoretical foundations will soon become a practical reality."[5]

In this era before the development of even our familiar jet engines, a key question for anyone envisioning space travel was what would propel a craft in the vacuum of space. In other words, what would the spacecraft push against to send it in the opposite direction? Tsiolkovsky solved the problem by using Newton's Third Law of Motion: for every action: there is an equal and opposite reaction. Tsiolkovsky cleverly illustrated the principle of rocket propulsion in space by imagining a boat stranded offshore with no oars or other means of propulsion. The hapless but clever sailor aboard notices a pile of large stones lying in the middle of the boat and begins to throw them overboard in the direction away from the shore. With each throw, the boat moves in the opposite direction—closer to shore—until it finally lands. The same principle applies to a rocket engine that jets exploding fuel out one end of a chamber, propelling the craft in the opposite direction, even in the vacuum of space. We might say that a spacecraft pushes against its own expelled fuel.

Tsiolkovsky's achievement is encapsulated in his rocket equation, which has three variables: a rocket's change in velocity starting from zero, its exhaust velocity, and the ratio of its initial to final masses. The equation represents a law of nature, akin to Newton's law of universal gravitation. By setting two of the three variables, the third becomes determined, leaving no room for alteration. The central concept is that the faster the fuel is expelled and the greater the fuel's mass relative to the rest of the rocket, the faster the rocket can travel.

For example, suppose you want a rocket to reach Earth's orbital velocity, approximately 8000 m/s. Suppose your rocket has a typical exhaust velocity of 3000 m/s. In that case, Tsiolkovsky's rocket equation shows that the ratio of the rocket's initial mass (including fuel) to its final mass (after the fuel is exhausted) must be about 14.5 to 1. This requirement greatly restricts the payload capacity and makes reaching orbit contingent on carrying a large amount of fuel.

This limitation led Tsiolkovsky to propose the concept of multistage rockets. The first stage burns through its fuel, detaches, and falls away. A second stage then ignites, taking over propulsion at a point when the rocket is significantly lighter due to the discarded first stage. This process can be repeated with a third or fourth stage, as necessary. By shedding weight sequentially, the multistage design maximizes efficiency, enabling the rocket to achieve the high velocities needed for space travel. The multistage rocket became the essential concept that enabled the space age. Tsiolkovsky was also the first to suggest that liquid hydrogen and oxygen, which can be better controlled than solid fuel and which would generate more propulsive force for an engine, would make a superior rocket fuel.

Consider, for example, the Saturn V, which sent astronauts to the Moon. To escape Earth's gravity, it needed to reach a velocity of about 11,200 m/s. To meet this requirement, the rocket's initial mass (including fuel) had to be approximately twenty-three times its final mass. A single-stage rocket could not accomplish this, so the Saturn V was constructed with three stages, discarding empty fuel tanks and engines at each stage to maximize efficiency and achieve the necessary speed.

Many of Tsiolkovsky's manuscripts long went unpublished. However, after his death, his daughter collected his writings and donated them to the Communist Party, which passed them on to the Russian Academy of Sciences. Many of these works have since been translated into English and are accessible online.[6]

When Soviet teams searched the Nazi rocket-building facilities at Peenemünde on the Baltic coast at the end of World War II, they discovered a German translation of a book by Tsiolkovsky in which "almost every page . . . was embellished with Wernher von Braun's comments and notes."[7] Von Braun had not only led the German rocket program but would also go on to head the American program that ultimately took astronauts to the Moon.

Tsiolkovsky also influenced another pioneer of rocketry, Hermann Oberth, shown in Figure 2.4. Oberth was a Hungarian-German whose interest in space flight was initially sparked by Verne's books, which he is said to have read repeatedly, as one often does with a childhood favorite. Oberth was such a precocious youngster that by the age of thirteen, he had already concluded that "A cannon is not good for spaceflight. It must be done with a rocket."[8]

Oberth's scientific career had a discouraging start when, in 1917, the Prussian Minister of War rejected his design for a long-range, liquid-propelled rocket. He went on to pursue a PhD at the University of Heidelberg, but the faculty refused to accept his dissertation on rocket design. Oberth paid to have it published privately as *The Rocket into Planetary Space*. Although the book was only eighty-seven pages long, space historian Frank Winter assesses that it "probably exerted more influence on the development of spaceflight than the work of either Tsiolkovsky or Goddard . . . [and] was the cornerstone of the Space Age."[9]

Figure 2.4 Hermann Oberth (fifth from right) and 18-year-old Wernher von Braun (second from right) in Berlin in 1930, examining Oberth's liquid-fueled rocket.
National Air and Space Museum

Oberth's most important scientific contribution may have been his demonstration that a rocket would become more efficient as the speed of its exhaust gases increased. Technically, he showed that propulsion efficiency—the ratio of the kinetic energy carried by the exhaust to the total heat energy released by the fuel—rises with exhaust velocity. In simpler terms, the faster the exhaust leaves the rocket, the more thrust is generated for the same amount of fuel. This principle is fundamental to rocket design: higher efficiency means that rockets can carry heavier payloads, travel farther, and support longer missions without requiring proportionally more fuel.

Oberth became a German citizen in 1940, and the following year he moved to the newly established German rocket factory at Peenemünde. There, he worked with von Braun, who would later say of him:

My honored mentor and teacher in rocketry, Professor Hermann Oberth . . . laid the scientific groundwork for the modern large liquid-fuel rocket and proved mathematically the potential capabilities of this type of rocket as a vehicle for extraterrestrial exploration.[10]

Figure 2.5   Robert H. Goddard (left) and team with his rocket.
National Museum of the United States Air Force

Robert Goddard, shown in Figure 2.5, was an American physicist who demonstrated through experimentation that Tsiolkovsky's rocket equation did indeed function in a vacuum. His childhood inspiration was H. G. Wells's *War of the Worlds*, serialized in the *Boston Post*. Like George Washington, Goddard had his own cherry-tree story. While trimming branches in the family tree, he drifted into a daydream about space travel that was so memorable that he later celebrated the date—October 19—as his "Anniversary Day."

In 1920, Goddard awoke to overnight fame. His scholarly report, *A Method of Reaching Extreme Altitudes*, written on behalf of the Smithsonian Institution, had caught the media's imagination.[11] The Smithsonian had made the report available to the public and issued a press release about it, which newspapers across the country picked up. Goddard had speculated that to demonstrate how far a rocket might travel, one containing flash powder could be sent to the Moon, where its explosion on impact might be visible from Earth. "New Rocket Devised by Prof. Goddard May Hit Face of the Moon," the *Boston Herald* headlined, and other newspapers around the world quickly followed suit. The news was irresistible, coming as it did at a time when the longest airplane flight had been a nonstop transatlantic crossing and when most Americans thought of a rocket as something to shoot off harmlessly on Independence Day. But Goddard's vision did not persuade everyone. In 1920, an overconfident editorial writer in *The New York Times*, reflecting the view at the time that in space, there would be nothing to push against, scolded, "Professor Goddard, with his 'chair' in Clark College . . . [who] does not know the relation of action to reaction, and of the need to have something better than a vacuum against which to react . . . knowledge ladled out daily in high schools." In an editorial published on July 17, 1969, one year short of a half-century later, as Apollo 11 headed for Tranquility Base on the Moon, the *Times* belatedly apologized to Goddard, admitting that he had "now definitely established a rocket can function in a vacuum as well as in an atmosphere."[12]

In 1936, the Smithsonian published Goddard's monograph, *Liquid-Propellant Rocket Development*. However, since it provided no practical engineering information, the report had little influence.[13] By this time, the Germans had been researching liquid-propelled rockets for several years. They had opened what would become their vast rocket research center at Peenemünde, which would soon employ thousands of engineers and technicians, along with large numbers of slave laborers. Goddard's rockets were designed to ascend as high vertically as possible. The German V-2 was intended to fly horizontally for hundreds of kilometers and to drop explosives on English cities. Even in those terms, the German rockets were far superior to Goddard's. In 1934, an early version of the V-2 reached an altitude of 3.8 km, compared to 0.6 km for Goddard's rocket.

The Smithsonian had supported Goddard's research during World War I. However, in contrast to the large investment that Germany made in rocketry, Goddard had to spend the final years of his career working in relative isolation at a small research station near Roswell, New Mexico, funded by industrialist Daniel Guggenheim. Intensely secretive, Goddard avoided interviews, with the *Boston Herald* noting, "His invention he will not discuss."[14] He worked with a small team of assistants, yet he never disclosed his overarching goals and he required them to sign agreements prohibiting the disclosure of any details of their work.

In an article titled "Pioneers of a New Age," published in the *New York Times* as Apollo 11 reached the Moon, Wernher von Braun summarized: "Goddard proved that liquid propellant rockets could be built and that they would perform as he and Tsiolkovsky had predicted." He credited Goddard with conducting "most of the basic research and development that made possible rockets such as the Saturn 5 [the rocket behind the Apollo program of crewed spaceflight]."[15]

American physicist and space science pioneer Herbert Friedman summed up the state of American rocketry at the end of World War II:

> The United States had no rocketry more sophisticated than the bazooka, a small antitank rocket launched by infantrymen from a tube resting on the shoulder, and the JATO (or Jet-Assisted Takeoff rocket), which was mounted on an aircraft to increase its lift on takeoff. The Germans, however, had made greater progress: they shocked the Western world with their monster rocket, the V-2, capable of launching a 2,000-pound explosive payload from the European continent to the British Isles and other targets.[16]

3

Vengeance Weapon to Vanguard: The Space Race Begins

Hitler's Rocket

Wernher von Braun (1912–1977) bridged two eras: that of the rocketry pioneers such as Robert Goddard and Hermann Oberth, and that of the generation that would carry astronauts to the Moon. Born in Wirsitz, Germany (now Wyrzysk, Poland), von Braun showed an early aptitude for mathematics and physics, which led him to study at the Berlin Institute of Technology. In 1934, he earned his doctorate in physics. The government classified his thesis on liquid-fuel rocket propulsion, demonstrating that even at this early date, Germany recognized the military potential of rocketry. The principles outlined in his thesis would serve as the foundation for developing the German V-2 rocket and those that followed, including the American Saturn V.

In 1937, von Braun began working at the new Peenemünde rocket center under the direction of military officer and engineer Walter Dornberger. Together, they secured funding for liquid-fuel rocket research and tested a rocket they named the A-2. It laid the groundwork for the much larger A-4, later renamed the V-2, or Vergeltungswaffe-2 (German for "Vengeance Weapon"), shown in Figure 3.1. This rocket was designed to avenge the "stab in the back" that Nazi mythology claimed led to Germany's defeat in World War I. The V-2 stood 14 m tall and carried a one-ton warhead. Its liquid-fuel engine allowed it to reach altitudes of 80 km and speeds above the sound barrier, making it the first long-range ballistic missile. (The term *ballistic* refers to a missile that has a powered launch with enough velocity to reach its designated target but once the engine burns out, continues on a trajectory governed by gravity and aerodynamics alone, much like an artillery shell. Ballistic missiles can achieve remarkable accuracy.)

Launched from sites in France beginning in 1944, the V-2 inflicted terror on civilians in London and Antwerp, but it had minimal military impact. Relative to its cost, which as a percentage of the German national budget

Sentinels in the Sky. James Lawrence Powell, Oxford University Press. © James Lawrence Powell (2026).
DOI: 10.1093/9780197842850.003.0003

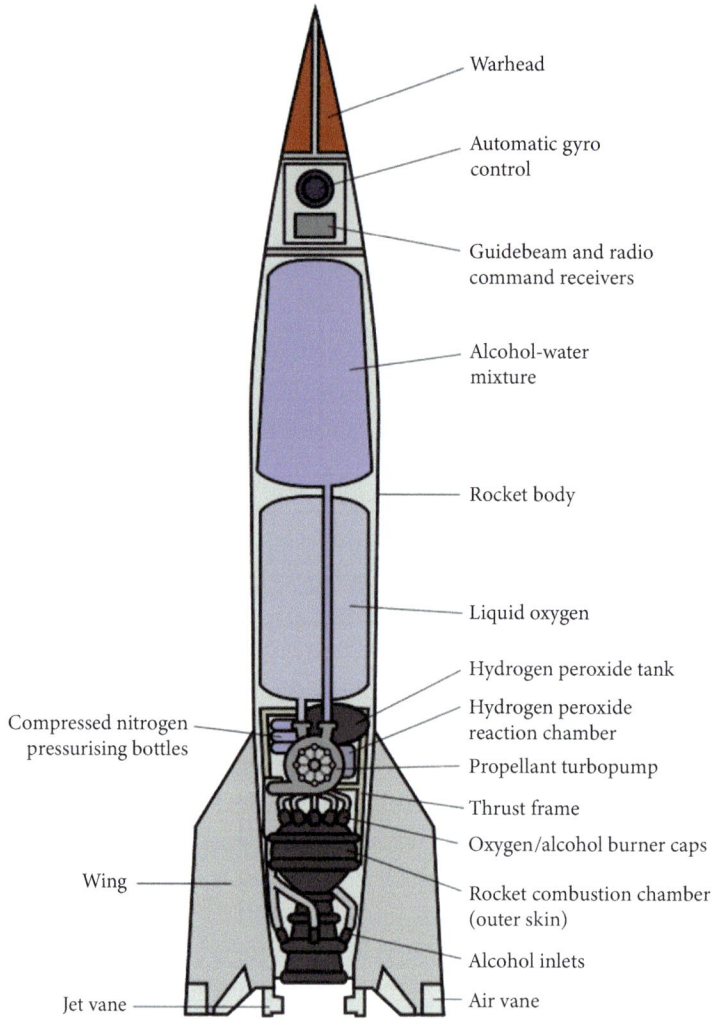

Warhead

Automatic gyro control

Guidebeam and radio command receivers

Alcohol-water mixture

Rocket body

Liquid oxygen

Hydrogen peroxide tank

Hydrogen peroxide reaction chamber

Compressed nitrogen pressurising bottles

Propellant turbopump

Thrust frame

Oxygen/alcohol burner caps

Wing

Rocket combustion chamber (outer skin)

Alcohol inlets

Jet vane

Air vane

Figure 3.1 German Vergeltungswaffe-2, or A-4—the "Vengeance Weapon." It was 14 m long.
Wikimedia Commons

was comparable to that of the U.S. Manhattan Project, the V-2 made only a negligible contribution to Germany's war effort.

As the war drew to a close, von Braun and his team fled west, hoping to surrender to the Americans rather than the Soviets. In a bold maneuver, they concealed vital technical documents and equipment in an abandoned mine in the Harz Mountains, ensuring that these materials would reach the U.S. military first. In May 1945, von Braun and his team surrendered to American forces in Bavaria, leading to their inclusion in Operation Paperclip, a U.S.

Figure 3.2 German rocket scientists at Ft. Bliss, Texas, in 1946. Wernher von Braun is seventh from the right, with a handkerchief in his breast pocket.
Wikimedia Commons

initiative designed to bring German scientists to America. It was given this harmless-sounding name because officers attached a paperclip to the dossiers of those who should be extradited to America.

Von Braun and 1,600 other German scientists—some with troubling pasts—were initially stationed at Fort Bliss, Texas (see Figure 3.2). Von Braun's past was itself troubling. He put men on the Moon—but he had also once built rockets for Hitler using slave labor. On one hand, he was an SS officer, which he needed to be to ensure his orders were obeyed. On the other hand, he was arrested by the Gestapo and jailed for two weeks without knowing the charges against him, suggesting that the German leaders were suspicious that he had insufficient loyalty to the Nazi cause.

Albert Speer, who served as Germany's Minister of Armaments and War Production from 1942 to 1945, may have spoken for both himself and von Braun when he made the following statement in a 1958 speech: "[T]he sober fact is that people, whether scientists or candlemakers, learn to live with such a situation. . . . The man living under dictatorship adjusts himself to business as usual, whether he likes it or not, because he must, in order to survive."

American Satellite

Under von Braun's leadership, on May 10, 1946, at the White Sands Proving Grounds in New Mexico, the site of the first atomic bomb test, the United States launched the first of the A-4 rockets that it had captured and brought to America. In contrast to the secrecy of the Soviet rocket program, the

U.S. Army invited mayors, military leaders, photographers, and reporters to attend. *Life Magazine* carried a five-page photo spread of the launch. The A-4 rocket rose straight up and then pitched to the north, reaching about 115 km in altitude and crashing 56 km downrange in the northern section of the range. By 1951, the United States had test-fired sixty-seven of the one hundred A-4 rockets it had brought to White Sands. Even though German rocketry ended with the war, both the U.S. and the Soviet rocket programs were direct descendants of von Braun's Peenemünde rocket.

In 1950, von Braun's team relocated to Redstone Arsenal in Huntsville, Alabama, marking the start of the U.S. Army's ballistic missile program. As technical director, von Braun led the development of the Redstone rocket, a short-range missile designed to carry a nuclear warhead. Adapted from the A-4 (switching back to civilian nomenclature), it significantly advanced the U.S. rocket program. The A-4 later evolved into the more powerful Jupiter-C, which underwent three test flights in 1956 and 1957. Though it carried a smaller payload than the V-2, it reached much higher altitudes, climbing to nearly 1,100 km.

Meanwhile, the U.S. Naval Research Laboratory, in collaboration with the newly established Jet Propulsion Laboratory in Pasadena, had been developing an experimental Viking rocket to launch a satellite named Vanguard. This initiated a rivalry between the two military branches to match that with the Soviet Union. Von Braun planned to launch a satellite in 1954, but the Pentagon refused permission. This decision reflected the military's priority; it sought to develop an Intercontinental Ballistic Missile (ICBM) capable of delivering a nuclear weapon rather than furthering space exploration. On September 20, 1956, to test the rocket's nose cone during reentry, a Jupiter-C lifted off from Cape Canaveral, which was later renamed the Kennedy Space Center, equipped with a fourth stage filled only with sand. The rocket had sufficient speed for the fourth stage to be released into orbit, but the Department of Defense prohibited the engineers from doing so. Had the fourth stage been deployed, the Army program would have beaten both the Soviets and the Navy into space by a year.

Soviet Rocketry

During the 1930s, like their German counterparts, Soviet researchers focused on developing rockets for military rather than scientific purposes. None of the rockets was more successful than the *Katyusha* (the Russian diminutive of Katherine), powered by a mixture of nitrocellulose and nitroglycerin.

It measured 1.8 m in length, had a range of about 5 km, and carried a 22-kg explosive warhead. During World War II, *Katyushas* were typically fired in barrages from ordinary trucks specially modified for the purpose. After firing a salvo, to evade counterattacks the trucks would speed to a new location. The *Katyusha* rocket motors emitted a howling shriek that terrified German troops, just as the V-1 Doodlebug rockets had frightened Londoners. Following Germany's invasion of the Soviet Union in 1941 in Operation Barbarossa, the *Katyusha* quickly demonstrated its value. The *Katyusha*, or a close derivative of it, has been used in many conflicts since, continuing to the present day.

Given the head start in rocketry that Tsiolkovsky had provided the USSR, one might have expected Soviet experts to create their own version of the V-2. This did not occur for two reasons. First, the Nazi invasion of the Soviet Union in June 1941 caught Stalin completely off guard. Although he received ample warning from Russian intelligence, he chose to trust Hitler rather than his own spies. With Stalin's delayed response, Soviet military command and control became paralyzed, resulting in the Germans advancing several hundred kilometers eastward within just three months to a line that stretched from near Leningrad in the north to Odessa on the Black Sea. The Soviets were engaged in a defensive war for survival, leaving no time for the extensive effort required to develop a V-2 equivalent.

Second, even if the Soviets had wished to start a military rocket program, perhaps at a safe distance from the conflict, by 1938 the top Soviet rocket experts had been imprisoned or executed, among the many victims of Stalin's Great Terror. In the second half of the 1930s, Stalin had unleashed an unprecedented program of state-sanctioned arrests, torture, show trials, forced confessions, and, for many, execution. The terror targeted the elite of Soviet society, including the upper ranks of the military, where thousands of officers—generals and high-ranking commanders among them—were accused of disloyalty, espionage, or collaboration with foreign enemies. Many were arrested, tried in secret courts, and sent to labor camps or executed. These purges enfeebled the Red Army, leaving it even less prepared to meet the Wehrmacht. In any event, the *Katyusha* was a far more lethal weapon than any Soviet V-2 could have been.

Stalin's paranoia consumed the Soviet scientific elite, including its rocket pioneers. Many were denounced for participating in "anti-Soviet" activities or conspiring with foreign powers. The Lysenko affair serves as a prime example: the prominent agronomist Trofim Lysenko's pseudoscientific theories, including a complete rejection of modern genetics, were endorsed by Stalin. As a result, thousands of mainstream biologists were silenced, arrested,

(a)   (b)

Figure 3.3 Sergei Korolev (left, 1934) and Valentin Glushko, leaders of the Soviet rocket program.
Wikimedia Commons

and, in many instances, put to death. Consequently, the biological sciences in the USSR suffered a generational setback.[1]

Given these developments, it is no surprise that those responsible for building rockets, which were inherently dangerous and could be directed wherever the rocket scientists wished, would attract Stalin's paranoia and become victims of his terror. Two early examples include the head of the institute that had developed the *Katyusha* and his deputy. They were shot in January 1938 after "confessing" and naming alleged collaborators.

By the mid-1930s, two engineers, Sergei Korolev and Valentin Glushko, shown in Figure 3.3, had emerged as the leaders of Soviet rocketry. However, by late 1937, the NKVD, or People's Commissariat for Internal Affairs, the Soviet Union's secret police agency, had secured statements denouncing both men as "wreckers," an all-purpose epithet that often led to a final bullet. Glushko was denounced by his colleagues and arrested in March 1938 for being an "enemy of the people," while four senior engineers at Korolev's institute condemned him on charges of being a member of an anti-Soviet organization. After spending a year in prison, Glushko received an eight-year sentence of hard labor. Korolev was arrested in June 1938 and, three months later, sentenced to ten years. Glushko was sent to a special prison for scientists and engineers, where he continued his rocketry research. In contrast, Korolev was moved to the notorious Kolyma death camp in far-eastern Siberia, part of the infamous gulag system described by Aleksandr Solzhenitsyn in *The Gulag Archipelago*. Then, in 1942, Korolev was transferred to a special aviation design bureau, and two years later, both men were freed to continue their work on rocketry. They would lead the Soviet space program for the next several decades. Had they died in the camps—a distinct possibility—Soviet rocketry would have faced years of delays.

MAD

Only a few months after Germany's surrender, the Soviet army dispatched a Special Brigade of rocket specialists, including Korolev and Glushko, to Germany to learn how to service and launch the captured A-4 rockets. They and others who would later lead the Soviet rocket program formed close personal connections during this stay in war-ravaged Germany. A year later, the team was transferred back to the Soviet Union to a newly established rocket center at Kapustin Yar, located 110 km southeast of Stalingrad. On October 18, 1947, the Soviets successfully launched the first of the A-4s they brought from Germany.

By this time, it had become apparent that the A-4 could be scaled up to carry much larger payloads and travel much longer distances—even deliver a weapon. On May 23, 1946, Stalin signed a special decree on "Matters of Rocket Weapons," so that both the United States and the USSR were embarked on a crash effort to build an Intercontinental Ballistic Missile (ICBM).[2] Korolev oversaw the design and construction of the Soviet rocket while Glushko's team built the engine.

Two key factors came together to make it possible for the United States and the USSR to launch artificial satellites just over a decade after World War II. First, the militaries of both nations needed rockets powerful enough to deliver nuclear warheads via ICBMs. These same rockets, designed for military purposes, were also capable of placing satellites into orbit. Second, visionary rocket scientists—such as Wernher von Braun, now in the United States, and Sergei Korolev in the USSR—had long dreamed of realizing Tsiolkovsky's early twentieth-century vision of launching artificial satellites. By the mid-1950s, both the military and the scientific communities had access to rockets powerful enough to turn that dream into reality.

Under Korolev's direction, the Soviet Union developed a series of increasingly advanced ballistic missiles. On February 2, 1956, the R-5M delivered a 1,300-kg nuclear warhead to a target 1,200 km away, where it detonated on impact. Fueled by liquid oxygen and alcohol, the R-5M became the first missile in history to deliver a live nuclear device, marking a pivotal moment in the development of strategic rocketry.

Analysis of the explosion revealed it had an energy equivalent to 80 kilotons of TNT, several times greater than that of the Hiroshima atomic bomb. This success prompted the Soviet Army to swiftly deploy the R-5M to its units in the field. Although the Cold War was already underway, this event could be seen as the beginning of the arms race. With such advances, humanity's fate would soon hinge on the untested strategy of "mutually assured destruction" (MAD): the notion that if two nations possessed sufficient nuclear weapons to inflict devastating damage on each other and if each could survive a nuclear first strike to launch its own second strike, then neither side would initiate a nuclear war. Surprisingly, considering human fallibility and the mutual distrust between nations, this strategy has worked.

The 1,200-km range of the R-5M was sufficient for a rocket launched from the USSR to reach targets in Europe but not the more distant American cities. In 1950, the Soviet government authorized a feasibility study for a true ICBM with a range of 5,000 to 10,000 km. By 1954, this initiative became a top priority. Three years later, Korolev had a successor rocket, the R-7 Semyorka, ready to launch a satellite (see Figure 3.4). Instead of stacking stages atop one another, it featured a cluster of rockets surrounding a central core. Each of the four strap-on boosters functioned as a separate rocket, equipped with its own fuel tanks and engine. The R-7 stood 34 m tall and was 10.3 m in diameter, weighing around 280 metric tons. The first R-7 exploded on the launch pad, and the next five attempts also failed, but the launch of Sputnik, as the world would soon learn, would be successful. After more than sixty years, descendants of the R-7, including the Soyuz, continue to be used for Russian space launches.

On August 21, 1957, TASS, the official Soviet news agency, announced the successful test of an ICBM, stating, "The results show that there is a possibility of launching rockets to any area in the world." American rocket scientists soon responded with the Air Force's Atlas ICBM, a descendant of the A-4, which had its first successful test flight on December 17, 1957. It boasted a range of about 14,000 km, enough to reach major Russian cities. The Atlas would also be repurposed for peaceful objectives through the U.S. Mercury program, launching astronauts into orbit.

The International Geophysical Year

As this intense competition in military rocketry unfolded in the 1950s, a group of American scientists sought to promote international cooperation in space. They modeled their efforts on the International Polar Years of

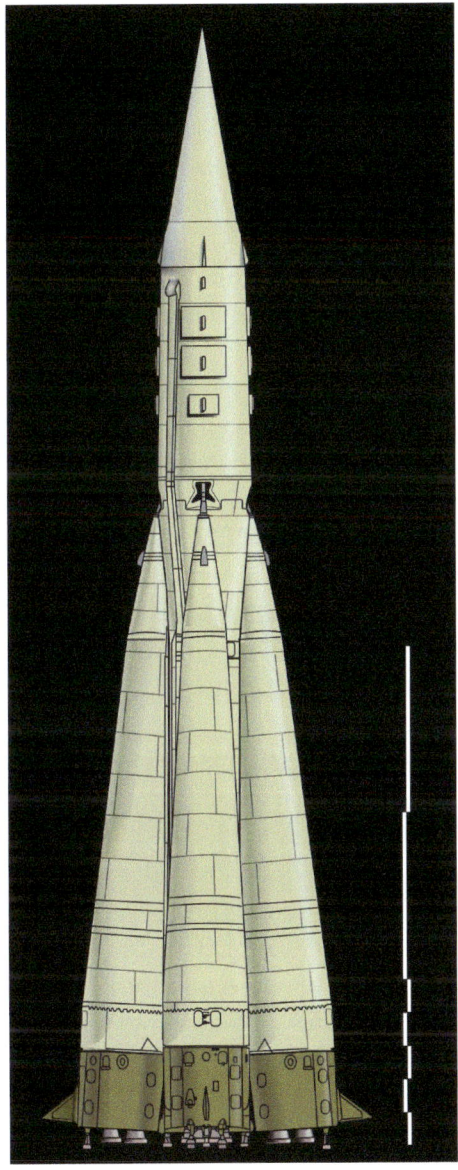

Figure 3.4 The Soviet R-7 rocket, the type that launched Sputnik. Note the cluster of smaller booster rockets around the base.
Wikimedia Commons courtesy of Heriberto Arribas Abato

1882–1883 and 1932–1933, when nations had collaborated to study the polar regions to enhance the accuracy of weather predictions and ensure the safety of air and sea transport. Their work eventually led to establishment of the World Meteorological Organization, which still operates today under the auspices of the United Nations. The American scientists then urged the International Council of Scientific Unions to approve a similar program. In 1952, the council announced the International Geophysical Year (IGY), which was to last from July 1, 1957 to December 31, 1958, the extra six months so that the program would overlap with the peak of the eleven-year sunspot cycle. The timing proved fortuitous as Stalin's death in 1953 cleared the way for the Soviet Union to participate in the IGY.

Space enthusiasts and rocket scientists, going back to Tsiolkovsky, recognized that a sufficiently powerful rocket could achieve orbital velocity and deploy an artificial satellite. Given the advances in rocketry by the early 1950s and the onset of the IGY, it was no surprise that both the United States and the USSR announced plans to launch a satellite as part of their IGY participation. On July 29, 1955, President Dwight Eisenhower's press secretary released a statement saying that the United States would launch "small unmanned earth-circling satellites" which would "for the first time in history enable scientists across the globe to make sustained observations in regions beyond the earth's atmosphere."[3] Less than a week later, at the Sixth Congress of the International Astronautical Federation in Copenhagen, Soviet physicist Leonid Sedov informed reporters that the USSR would soon place an artificial satellite into orbit.

Eisenhower's announcement prompted the Soviet government to instruct Korolev to transition from military rockets to the design, construction, and launch of a Soviet satellite. His initial effort resulted in a scientifically oriented device weighing three thousand pounds, called Object D, but it proved too heavy for the R-7 to lift. A year later, Korolev proposed a much simpler and lighter satellite, which the USSR Council of Ministers approved on February 15, 1957. It was named Object PS, short for *Prosteishii Sputnik*, meaning "simplest satellite." Sputnik translates to "fellow traveler." Just how far the Soviets had traveled, the Americans were about to learn.

4

Sputnik

4.1 Sputnik Launch

As a result of big intensive work by scientific-research institutes and design bureaus, the first-in-the-world artificial satellite has been created. On October 4, 1957, the first satellite was successfully launched in the USSR. . . . The successful launch of the first man-made earth satellite makes a most great contribution to the treasure- house of world science and culture. Artificial earth satellites will pave the way to interplanetary travel, and our contemporaries are apparently destined to witness how the freed and conscientious labor of the people of the new socialist society makes the most daring dreams of mankind a reality.

—TASS Announcement, in *Pravda*, October 5, 1957[1]

Fellow Traveler

The world learned that the Soviets were the first in space through the above announcement by TASS, the official Soviet news agency. Although the Soviets had made no secret of their plan to launch a satellite—quite the opposite, they had stated it on numerous occasions—the initial response of many Americans to knowledge of Sputnik overhead was disbelief: this was merely a trick, one of the ridiculous "firsts" that the Soviets had falsely claimed, which even included baseball, the quintessential American game. But skepticism could barely survive the weekend; then came anger and the need to find someone or something to blame (Figure 4.1).

Some noted that since the Soviets had obtained American atomic bomb secrets through espionage, why not those of rocketry as well? The trouble was, as one person commented, "Those Russian spies must be really good:

Sentinels in the Sky. James Lawrence Powell, Oxford University Press. © James Lawrence Powell (2026).
DOI: 10.1093/9780197842850.003.0004

Figure 4.1 Replica of Sputnik.
U.S. Air and Space Museum

they stole a secret we didn't even have yet."[2] Others claimed that Soviet satellites were designed to spy on the United States, much as the United States had been spying on the USSR through the by-then-not-so-secret U-2 program of high-altitude reconnaissance flights. The ill-focused anger grew worse as President Eisenhower downplayed Sputnik's significance and it became clear that the United States did not have a ready response. Some began to predict further Soviet achievements on the horizon, one authority saying, "I would not be surprised if the Russians reached the Moon within a week."[3] This prediction was not so far-fetched, as just months later the Soviets did send the probe Luna to crash into the Moon. It seemed reasonable to assume that Sputnik must be doing something up there rather than monotonously beeping—perhaps it was collecting and transmitting vital information back to the USSR.

Influential newspaper columnist Stewart Alsop wrote on October 13 that "There is a mounting body of evidence, taken very seriously within the Washington intelligence community, that the Soviet satellite is not blind ... Sputnik has eyes to see." If the Soviets could achieve orbital velocity with a small metal ball, it was not too much to imagine that they could use a satellite to deliver a nuclear warhead to any point on Earth. The director of the Smithsonian Astrophysical Laboratory speculated that the Soviets planned to detonate a hydrogen bomb on the Moon, in a colossal version of Goddard's old idea. One of the most astonishing reactions came on October 9 from President Eisenhower himself. The Russians had managed to be the first in space, he said, because "they had captured all the German scientists" at Peenemünde. In fact, just the opposite was true. Von Braun and his team were in Alabama, not the USSR. Moreover, it was the Eisenhower administration that had stifled their work in favor of the so-far unsuccessful Vanguard program and its military focus.

The Soviets had previously announced that Sputnik would be visible to the naked eye. However, its visibility turned out to be just at the limit of detection, making it likely that some, but far from all, would be able to see it. Nevertheless, many were convinced they had. What most had seen instead was the much larger and 100 times more visible R-7 booster, which was itself in orbit as it released the satellite. The booster remained aloft until December 2 before burning up in the atmosphere, while Sputnik lasted until January 4, 1958.

Yet there was another way to confirm the orbiting Sputnik: through its radio transmissions.[4] To track the satellites it planned to launch, the United States had built a worldwide network of ground stations called Minitrack, utilizing the 108-MHz frequency established by the IGY. Minitrack went into operation on October 1, three days before the launch of Sputnik. However, as the Soviets had said they would, Sputnik broadcast not at 180 MHz but at 20 and 40 MHz, frequencies that Minitrack could not receive. Nevertheless, these bands were used by amateur (ham) radio operators worldwide, who soon picked up Sputnik's signal and confirmed that it was in orbit. The Minitrack operators quickly followed suit, adjusting their equipment so that they, too, could receive Sputnik's signal.

President Eisenhower gave a televised address to the nation on November 7, by which time the Soviets had already launched Sputnik II with the doomed dog Laika aboard. He announced that James R. Killian Jr., president of MIT, would serve as the first White House science adviser. Ike also said, "One of our greatest and most glaring deficiencies is the failure of us in this country to give high enough priority to scientific education and to the place of science in our national life."[5] This statement identified a "science education gap" to go along with the alleged "missile gap" that candidate John F. Kennedy would highlight in the upcoming 1960 presidential election. In response, Congress passed the National Defense Education Act of

1958, which provided an unprecedented $1 billion in funding for education at all levels, with a focus on science and technology. This bill included financial aid for students pursuing higher education in science, technology, engineering, and math (STEM) fields, resulting in a significant increase in the number of Americans earning STEM PhDs. That year, the government also established the National Aeronautics and Space Administration (NASA) to concentrate on space research and exploration. These were long-range efforts, but the public still wanted to know when the United States would loft its own satellite.

Army vs. Navy

At the time the news of Sputnik broke, Secretary of Defense-designate Neil H. McElroy happened to be visiting the Redstone Arsenal in Alabama, home to the Army Ballistic Missile Agency (ABMA). When von Braun heard the news that the Soviets were the first, he responded: "We knew they were going to do it! [The Navy's] Vanguard will never make it. We [the Army] have the hardware on the shelf. For God's sake, turn us loose and let us do something. We can put up a satellite in sixty days, Mr. McElroy! Just give us a green light and sixty days."[6] Upon reflection, von Braun and Major General J. B. Medaris, who oversaw the ABMA, revised the time estimate to a more feasible yet still rapid ninety days. They made this bold prediction because, despite the Defense Department's order to stand down, von Braun's team had never stopped working to prepare the Jupiter C to launch a U.S. satellite. On November 8, five days after Eisenhower's speech, the Department of Defense made it official:

> The Secretary of Defense today directed the Department of the Army to proceed with launching an earth satellite using a modified Jupiter C. This program will supplement the Vanguard project to place an Earth satellite into orbit around the Earth in connection with IGY. All test firings of Vanguard have met with success [a claim soon to be contradicted] and there is every reason to believe Vanguard will meet its schedule to launch later this year a fully instrumented scientific satellite. The decision to proceed with the additional program was made to provide a second means of putting into orbit, as part of the IGY program, a satellite which will carry audio transmitters compatible with Minitrack ground stations and scientific instruments selected by the National Academy of Sciences.[7]

The Vanguard and Army programs now advanced in parallel, competing to see which would be the first to launch an American satellite. On October

11, 1957, the White House announced that a test flight of the Vanguard rocket carrying a satellite would take place in December. However, the press and the public viewed this effort as more of a genuine attempt to respond to Sputnik than merely a test. After several delays, at 11:44:55 A.M. on December 6, 1957, the rocket ignited, ascended a few feet into the air, and then collapsed under its own weight. The fuel tanks ruptured and exploded, destroying the rocket and severely damaging the launch pad. The small satellite-to-be escaped the flames, rolled into nearby bushes, and began to beep pitifully. This "four-inch" flight, which the *London Daily Herald* dubbed a "flopnik," created an opportunity for von Braun and the ABMA program to succeed where the Navy's Vanguard had faltered.

On December 20, 1957, preflight tests commenced on the Army's Jupiter-C rocket, led by Peenemünde alumnus Kurt Debus, later the first director of Kennedy Space Center. To further complicate matters, the Jupiter-C, which had originally begun as the Redstone, was now renamed Juno, the sister and consort of Jupiter. By mid-January, the Navy's replacement for the destroyed Vanguard rocket had nearly launched twice, only for the countdown to have to be halted. This placed von Braun's team in the lead, and by January 28, they were ready. Then came the seemingly unavoidable inclement weather and delays. The next Vanguard launch was scheduled for February 3, and unless von Braun's Army rocket could lift off before then, it risked tumbling to an ignominious third place in the race to orbit a satellite. Adding to the pressure, the ninety-day delivery period that Medaris and von Braun had promised would end on February 6, 1958.

Von Braun Delivers

On January 12, 1958, an Air Force meteorologist forecast that the winds at the launch site would weaken, creating an opportunity for lift-off. At 10:48 EST, Juno was on its way. It successfully passed through the high-altitude jet stream, only identified in the 1940s. By the seven-minute mark, the second, third, and fourth stages had propelled Juno to orbital speed, from where it released the satellite, which would be named Explorer. The first challenge was the same one that had immediately followed Sputnik: how to confirm that Juno was in orbit rather than having exploded or gone off course. After a nail-biting 100 minutes, the Jet Propulsion Laboratory confirmed that Explorer's signal indicated it was indeed in orbit.[8] At a 2:00 p.m. press conference at the National Academy of Sciences the following day, von Braun and other officials lifted a model of the satellite above their heads in celebration. It would be called Explorer 1 (Figure 4.2).

Figure 4.2 Left to right: Jet Propulsion Laboratory Director Dr. James Pickering, Dr. James Van Allen of the State University of Iowa, and Dr. Wernher von Braun holding a model of Explorer aloft.
NASA

James Van Allen's presence in the press conference photo underscored that, unlike Sputnik I, the U.S. satellite was launched not merely as a technological feat, but to serve the scientific goals set forth by the International Geophysical Year. Dr. James Pickering, also shown, was in charge of constructing Explorer I, while Dr. Van Allen oversaw its scientific payload. One of the instruments aboard detected cosmic rays: high-energy protons or atomic nuclei traveling through space at nearly the speed of light. The detector yielded puzzling results: at times it recorded the expected number of cosmic ray hits, but at other times it detected none. The zero counts occurred when the satellite was at an altitude exceeding 2,000 km. Results obtained from Explorer 3, launched two months later (after Explorer 2 had failed), showed that the zero counts were due to the counter receiving so many cosmic rays that it became overwhelmed and could not record any. Van Allen and his team at the University of Iowa deduced that this was a result of a belt of charged particles trapped in space by Earth's magnetic field. They were named the Van Allen radiation belts in honor of their discoverer.

On March 17, 1958, a successful launch placed the Vanguard satellite into orbit, the first to be powered by the Sun. The Vanguard program conducted a total of eleven satellite launches, succeeding with three of them. The Vanguard 1 satellite remains in orbit, along with the third stage of its launch rocket. It is the oldest human-made object in space. The satellite's prolonged lifespan has provided invaluable data for understanding atmospheric drag and other long-term orbital mechanics.

Both the Vanguard and Explorer programs were soon absorbed into the newly established NASA, bringing an end to the Army–Navy rivalry and laying the groundwork for a unified era of scientific satellite missions.

Although Sputnik 1 was not designed with scientific objectives in mind, it nonetheless made a groundbreaking contribution: by tracking its orbit, scientists were able to exploit the predictable motion of fast-moving objects in space—a principle that would become foundational to satellite technology and space science.

PART II

ECHOES IN THE SKY

5
The Doppler Effect

Tsiolkovsky's rocket equation established an implacable limit: larger satellites require larger rockets and more fuel. This was why Korolev had to postpone the launch of the satellite that would become Sputnik 3, which weighed nearly 3,000 pounds, and opted instead to begin with the 184-pound Sputnik 1. As a result, the small sphere had limited space for the scientific instruments that would later become a standard feature of satellite launches. Nevertheless, Sputnik 1 did provide valuable scientific information. At the apogee of its elliptical orbit, when farthest from Earth, it reached an altitude of 939 km. At perigee, when it was closest to Earth, it was 215 km away. This indicated that Sputnik 1 was designed to operate entirely within the ionosphere, the region of the atmosphere ranging from about 48 km to approximately 965 km (see Figure 13.4). The ionosphere gets its name from the cosmic rays and solar ultraviolet radiation within it that strike neutral atoms of atmospheric gases, converting them into electrically charged ions. This process allows the ionosphere to propagate radio waves over vast distances. Although many scientists had predicted the existence of the ionosphere, it was Sir Edward Appleton, a British physicist, who first confirmed it through experiments in the 1920s. He transmitted radio waves vertically into the atmosphere and observed their reflection off an upper layer back to Earth's surface. For that discovery, Appleton received the 1947 Nobel Prize in Physics. Sputnik 1's signal confirmed what physicists already understood: the ionosphere transmits radio waves. By examining the details of the radio transmission, Soviet scientists could garner vital information about the nature of the ionosphere. Ingenious engineers also devised a way to use Sputnik's fan, which turned on and off at specific temperatures, to estimate the ionosphere's temperature at various altitudes. However, Sputnik's most significant discovery lay in its fluctuating radio signal—a finding that continues to influence daily life.

Sentinels in the Sky. James Lawrence Powell, Oxford University Press. © James Lawrence Powell (2026).
DOI: 10.1093/9780197842850.003.0005

The Sound of Speed

We are all familiar with the Doppler Effect, even if we may not recognize the name. Through our own experiences and in films and on television, we have heard a car or train approaching, its sound rising in pitch and then falling again as it passes. Austrian physicist Christian Doppler first described this effect in 1842. The rise in pitch occurs because, as the sound source approaches, the sound waves are compressed closer together. As illustrated in Figure 5.1, as the waves bunch up, their frequency and pitch rise. As the source moves past and away, the opposite occurs, and the frequency and pitch decline (not shown).

The formula for the Doppler Effect is relatively simple for sound waves and allows the speed of a moving source to be calculated from the change in frequency.

Speed of source = (Speed of wave × Observed frequency)
÷ Original frequency

For light, especially at high speeds, the predictions of Einstein's special relativity replace those of classical physics. In this case, if the observed light frequency is half the emitted frequency, relativity shows that the source is receding at about 60 percent of the speed of light—rather than 50 percent,

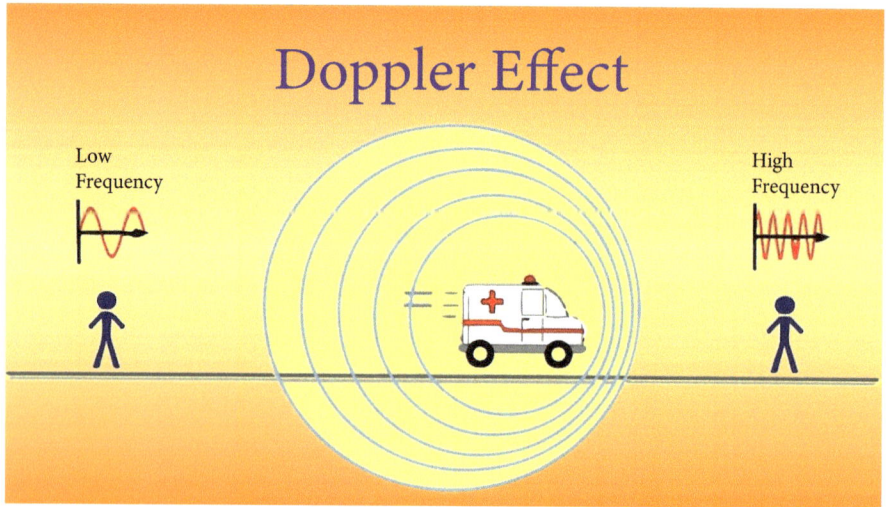

Figure 5.1 The Doppler Effect produced by a speeding ambulance moving from left to right, showing the bunching up of sound waves and the rise in frequency.
Wikimedia Commons

as classical Doppler theory would suggest The first to apply the Doppler Effect to Sputnik were William H. Guier and George C. Weiffenbach, who, at the time of its launch, worked at the Applied Physics Laboratory of Johns Hopkins University.

Recall that Sputnik transmitted signals on two frequencies, one at 20 megahertz. Using that frequency, Guier and Weiffenbach measured Sputnik's Doppler shift, which turned out to correspond to a speed of about 8,000 m/s—the velocity required for orbit. This demonstrated that Sputnik was not a hoax but a genuine Soviet satellite. The two scientists used the Doppler shift to determine that Sputnik's elliptical orbit allowed it to pass over virtually all of Earth's surface except for the polar regions, completing one orbit every 96 minutes. Many scientists and radio enthusiasts tracked Sputnik, but only Guier and Weiffenbach applied the Doppler shift to ascertain its speed and orbital path.[1]

An object traveling at a constant speed exhibits a constant Doppler shift. However, Sputnik's shift changed from moment to moment, indicating that it was both accelerating and decelerating. Through detailed analysis, Guier and Weiffenbach discovered that from one set of Doppler-shift data, "a complete set of orbit parameters for a near-Earth satellite could be inferred to useful accuracy."[2] Thus, from the Doppler shift alone, a ground station could accurately locate a satellite's position in space and measure its speed.

Inverting the Calculation

The breakthrough came on March 17, 1958. Frank McClure, chairman of the Applied Physics Laboratory, asked Guier and Weiffenbach whether they could "invert" the calculation: that is, could the precisely known orbit of a satellite and its Doppler shift be used to determine the geographic location of the receiving station on the ground? The two researchers immediately began working on the problem and found that "[t]he very first simulations indicated great accuracy—unbelievable accuracy!" The U.S. Navy was vitally interested in this subject since its *Polaris* nuclear submarine, which on short notice might need to launch a nuclear missile from anywhere in the ocean, had to know its precise location at the time of launch. More broadly and far more significantly, this "inverted solution" was the key insight that led to development of satellite-based navigation systems.

After the initial discovery by Guier and Weiffenbach, scientists at Johns Hopkins rushed to develop a global, space-based location system. The result was the U.S. Navy's Transit system, a constellation of ten satellites at an altitude of 960 km, enough to ensure worldwide coverage with built-in redundancy. The Transit satellites were equipped with solar panels that deployed in space, providing a long-lasting power source and helping to stabilize their orbits. Transit became operational in 1964 and was used primarily by *Polaris* submarines and other naval vessels, as well as by land surveyors, who averaged individual measurements to improve accuracy. Transit remained in operation until 1991, when it was succeeded by the full deployment of what we now know as the U.S. Global Positioning System, or GPS. Satellite navigation provided several immediate advantages, one of which was the ability to deliver a continuous signal, whereas Transit could only offer about one reading per hour.

The U.S. Global Positioning System

As we have seen, the idea of launching an artificial satellite originated with Tsiolkovsky and the pioneers of rocketry. The germ of the concept of using a constellation of satellites to determine the location of a receiving station on Earth, as well as a number of other fruitful ideas, we may well owe to the brilliant scientist and science-fiction writer Arthur C. Clarke. In August 1956—one year before Sputnik—Clarke wrote to a friend:

> As you may know, my main interest in this subject is in the use of satellite relays, which I think may revolutionise the pattern of world communications. To the best of my knowledge, I was the first to suggest this possibility (see "Extraterrestrial Relays," Wireless World, October 45). My general conclusions are that perhaps in 30 years the orbital relay system may take over all the functions of existing surface networks and provide others quite impossible today. For example, the three stations in the 24-hour orbit could provide not only an interference and censorship-free global I V service for the same power as a single modern transmitter, but could also make possible a position-finding grid whereby anyone on earth could locate himself by means of a couple of dials on an instrument about the size of a watch. (A development of Decca and transistorisation.) It might even make possible worldwide person-to-person radio with automatic dialing. Thus no-one on the planet need ever get lost or become out of touch with the community, unless he wanted to be. I'm still thinking about the social consequences of this!

> But as for details of frequencies and powers, I'll have to leave that to the experts to work out; I'll get on with my science fiction and wait to say, "I told you so!"[3]

In the mid-1960s, Roy Anderson, a consulting engineer at General Electric, proposed what would become our modern global navigational system.

Instead of utilizing the Doppler shift, he suggested measuring the distance between a satellite and a receiver by calculating the time it takes for radio waves, which travel at the speed of light (299,792,458 m/s in a vacuum). For instance, if a signal from a single satellite takes exactly 0.1 seconds to arrive, we could envision the receiver as being located on the surface of an imaginary sphere 29,979.2 km from the satellite. The signals from two additional satellites create three spheres that intersect at two possible points, one of which is clearly incorrect, such as being deep in space or underground. The remaining issue is that the receiver's internal clock lacks the precision of the atomic clocks aboard the satellites, introducing a small error in measuring the signal travel time. But small is relative: an error of just 1 ns—a billionth of a second—translates to a positional error of about 30 cm on the ground. The signal from a fourth satellite is crucial because it eliminates uncertainty in the receiver's position and allows for accurate synchronization of time, which is essential for pinpointing precise locations.

Anderson recognized that Transit's limited number of satellites constrained its accuracy. In 1964, he proposed a constellation of twenty-four satellites, each with a six-hour orbit. In the resulting Space Surveillance System, transmitters on Earth broadcast signals into space, which were reflected when a satellite crossed the broadcast beam, providing the distance information needed to locate the ground station. Anderson's key concept was that the system should be "passive": instead of requiring the receiver to contact the satellite, the system would be "always on," providing a continuous signal. This would allow any number of users to receive the satellite's signal and determine their positions simultaneously, just as we can all receive the same radio, television, and satellite transmissions today. Such continuity would be essential for military users, who need accurate and instantaneous location measurement. The U.S. Department of Defense adopted Anderson's suggestion in 1973, and the constellation of twenty-foursatellites went into operation in 1993.

Modern smartphones and receivers are impressively accurate, generally able to determine location within approximately 5 m under normal conditions. This accuracy further improves when additional information from cell towers and Wi-Fi networks is included. However, the true potential for extreme accuracy resides in another aspect of satellite signals, making them invaluable for scientific and technical applications.

Radio, television, and satellite signals all use electromagnetic waves—called carrier waves—to send information through space. A carrier wave is a continuous electromagnetic wave that can be modified, or modulated, to carry data such as sound, images, or digital signals. By altering the wave's amplitude, frequency, or phase, information is encoded onto the carrier. The receiving device—like a radio, TV, or satellite dish—then decodes this modulation to retrieve the original content. This method allows information to travel over long distances, including through the atmosphere and even across space.

Navigation satellites emit two such carrier waves: the L1 band, designed for civilian use, and the L2 band, intended for military and high-precision applications. When both bands are used together, receivers can account for atmospheric disturbances—such as signal delays caused by the ionosphere—that can degrade accuracy. By using both carrier signals, positioning accuracy can improve by a factor of up to 1,000, reducing the margin of error to just a few centimeters.

This enhanced precision has transformed satellite technology from a tool used primarily for navigation into a vital resource for scientific research. Applications include tracking tectonic plate movements, measuring ice sheet dynamics, optimizing agricultural practices through precision farming, and monitoring global environmental change. Essentially, the ability to harness the dual carrier waves has opened new frontiers in how we understand and interact with the world.

The Global Navigation Satellite System

Given the significance of satellite systems for navigation and determining a receiver's position on Earth, it is not surprising that many nations have developed their own systems, as depicted in Figure 5.2. The American Global Positioning System and the Soviet GLONASS both began initial operations in 1993 and achieved full operational capability in 1995. GLONASS fell into disarray after the dissolution of the USSR but was restored to full capacity in 2011. However, the unprovoked Russian invasion of Ukraine has led to proposals for withdrawing international support for GLONASS. The Galileo program of the European Union started operations in 2016 and reached full capability with twenty-four satellites by 2024. China launched its BeiDou system in phases, achieving full operation by 2020. Collectively, these four programs are referred to as the Global Navigation Satellite System, or GNSS. India and Japan have initiated their own more localized programs.

Each GNSS system maintains a constellation of satellites, balancing military needs with civilian convenience. The latest GNSS receivers can use signals from any combination of the four systems and may track twenty to thirty satellites simultaneously, resulting in the highest possible accuracy.

While these systems share similarities, they differ in technical implementation and strategic focus. GPS and Galileo satellites operate on the same frequencies but are distinguished by unique coding. GLONASS initially used

4 GNSS CONSTELLATIONS

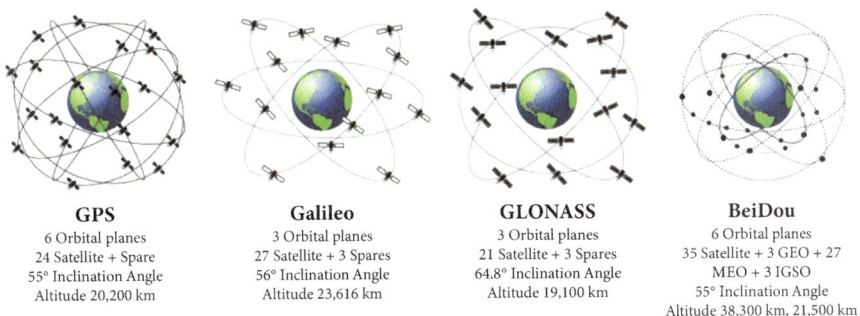

GPS	Galileo	GLONASS	BeiDou
6 Orbital planes	3 Orbital planes	3 Orbital planes	6 Orbital planes
24 Satellite + Spare	27 Satellite + 3 Spares	21 Satellite + 3 Spares	35 Satellite + 3 GEO + 27
55° Inclination Angle	56° Inclination Angle	64.8° Inclination Angle	MEO + 3 IGSO
Altitude 20,200 km	Altitude 23,616 km	Altitude 19,100 km	55° Inclination Angle
			Altitude 38,300 km, 21,500 km

Figure 5.2 Global Navigation Satellite System (GNSS) in 2021.

Space News/Penn State

separate frequencies but is transitioning to the GPS-Galileo coding approach. BeiDou stands out with its hybrid system, combining medium Earth and geostationary orbits (see Figure 9.5), which enhances regional coverage in the Asia-Pacific. Galileo prioritizes civilian applications and boasts the highest signal accuracy. GLONASS remains closely aligned with Russian strategic interests, whereas BeiDou functions not only as a navigation system but also as a key element of China's broader geopolitical and technological ambitions.

Even when global navigation systems are using signals from four or more satellites, they must still correct for several sources of error. Two key examples are (1) ionospheric delay—satellite signals slow down as they pass through the ionosphere, where they interact with charged particles, including free electrons and ions; and (2) multipath interference—signals can reflect off surfaces such as buildings, water, or terrain, causing them to travel longer paths and thus introduce timing errors. This effect is especially pronounced in urban environments, where numerous reflective surfaces create complex signal paths. The Galileo system addresses this issue by filtering out reflected signals based on their polarization—the orientation of the electromagnetic wave's vibrations.

To avoid confusion, when writing about the satellite program of a specific country or partnership, we will use the appropriate name, such as GPS or Galileo. Conversely, when referring to satellite technology in general, we will use GNSS.

6

Satellite Navigation

Military Uses

The Department of Defense initially developed the United States Global Positioning System as a tool for military navigation and targeting; it became operational for limited use in the 1980s and was fully deployed by 1993.

Accurate navigation has always been critical in warfare, and GPS provided a revolutionary improvement. With precise positioning available anywhere in the world and the system "always on," military forces could achieve unprecedented accuracy and coordination. Soldiers, vehicles, aircraft, and naval vessels could determine their exact positions in real time, reducing the risk of navigational errors and enabling forces to traverse challenging terrain with confidence.

Before satellite technology, guided missiles were steered using inertial navigation systems (INS), which are self-contained methods that calculate position, velocity, and orientation by measuring acceleration and rotation. These systems use accelerometers and gyroscopes to track changes from a known starting point, integrating their readings over time. Although INS is autonomous and dependable, it is prone to cumulative drift: small errors in measurement gradually accumulate, leading to increasing inaccuracy. To maintain precision, periodic corrections from external references are necessary. One of the first military uses of satellite technology was to correct INS measurements for ballistic missiles traveling at high speeds. These missiles must maintain an accurate trajectory to reach their targets, which are often located thousands of kilometers away, potentially on the opposite side of the world. Using satellite correction enables the inertial navigation system to stay aligned despite the extreme dynamics, high accelerations, and varying atmospheric conditions that may worsen drift errors. The Gulf War demonstrated GPS's capabilities, as bombs guided with meter-level precision struck targets with minimal collateral damage. Many will remember videos showing a smart bomb entering through a window or down a chimney.

In contrast, submarines, which often rely primarily on inertial navigation, obviously are less affected by high-speed dynamics and can operate effectively

Sentinels in the Sky. James Lawrence Powell, Oxford University Press. © James Lawrence Powell (2026).
DOI: 10.1093/9780197842850.003.0006

for extended periods using INS alone. In the early era of missile submarines, inertial navigation systems were the primary method for navigation and targeting. These systems allowed submarines to operate stealthily and autonomously for extended periods, as they did not rely on external signals that could be intercepted or traced, thereby preserving the vessel's hidden position. Today's submarines use satellites to periodically update and recalibrate their INS, ensuring operational autonomy when satellite signals are unavailable (e.g., underwater or in adversarial environments where signals might be jammed). Meanwhile, satellite updates help minimize drift errors and provide reliable long-term accuracy.

When Polaris submarines began patrolling, satellite navigation was in its early stages. One of the first satellite-based navigation aids, as noted, was the Transit system, which became operational in 1964. Transit allowed submarines to determine their positions by receiving signals from satellites in polar orbits. By analyzing the Doppler shift of these signals, they could accurately calculate their latitude and longitude. By the time GPS became fully operational, it was integrated into the Ohio-class submarines, which replaced the Polaris fleet.

Satellites have revolutionized military logistics by tracking troops without disclosing their positions. Additionally, satellite technology improves search-and-rescue missions by enabling accurate location tracking, thus saving lives during efforts to recover downed pilots or stranded personnel. Modern conflicts, such as the Russia-Ukraine War, have seen a dramatic rise in military drones, which depend on satellite signals as their primary means of navigation.

Civilian Uses

Though designed for war, GPS soon proved indispensable in civilian life. However, at first, civilian signals were intentionally degraded under a policy that limited accuracy to about 100 meters, so as to prevent potential adversaries from using the system for precision targeting or other military applications. The encrypted military signal, by contrast, retained full accuracy.

Aviation and shipping embraced GPS early, recognizing its benefits for safer routes, reduced fuel consumption, and fewer delays. Ships used GPS for accurate navigation across oceans, which was especially beneficial during adverse weather conditions when traditional methods, such as celestial navigation, became unreliable.

Surveyors and cartographers quickly embraced satellite technology for land measurements and mapping, facilitating faster and more precise results than traditional methods. Emergency responders began using satellite signals to locate accident scenes, coordinate disaster responses, and guide ambulances or fire trucks to locations with greater speed and accuracy. By the late 1980s, satellite-based devices emerged in consumer markets for outdoor activities such as hiking, sailing, and geocaching. Although these devices were expensive and bulky compared to modern receivers, they demonstrated the Global Navigation Satellite System's (GNSS's) potential to improve personal navigation. Satellite signals became essential for accurate timekeeping in telecommunications, banking, and other industries requiring synchronized operations.

By the early 1980s, satellite technology was rapidly advancing, ushering in a new era of precision in navigation, communication, and global surveillance. It appeared that satellites would revolutionize how humans interacted with the world, offering unprecedented capabilities and a brighter, safer future. However, this optimism was soon clouded by a tragic event that underscored the profound consequences of human error and geopolitical tension in an era that had become increasingly reliant on technology.

KAL 007

On September 1, 1983, Korean Air Lines Flight 007 (KAL 007), a Boeing 747 with 269 passengers and crew aboard, departed New York bound for Seoul. Unbeknownst to the crew, the aircraft's autopilot failed to correct a navigational drift, causing the plane to veer steadily off course. By the time it entered Soviet airspace over Sakhalin Island, north of Japan, it had strayed nearly 580 kilometers from its intended route   deep into a sensitive military zone. Mistaking it for a hostile intruder, Soviet air defenses launched a fighter jet that intercepted and ultimately shot down the aircraft, killing all on board.

The tragedy became one of the tensest moments of the Cold War, as Soviet authorities asserted that KAL 007 was a U.S. spy plane. The underlying cause of the incident was the aircraft's dependence on inertial navigation, which, without the benefit of satellite-based corrections—then available only to the military—was susceptible to gradual, undetected drift. Over time, these small errors accumulated, leading the plane far off course without the crew realizing it.

In response to the disaster, President Ronald Reagan declared that GPS should be made available for civilian use, thereby ensuring that no aircraft would be lost due to navigational error. However, to protect military interests, the civilian GPS signal was initially intentionally degraded. This restriction remained in place until the year 2000, when President Bill Clinton lifted it, granting the public full access to GPS accuracy.

How Location by Satellite Works

As illustrated in Figure 6.1, global positioning works through a series of lightning-fast steps, of which we are entirely unaware:

1. **Satellite Time Synchronization**—GPS satellites are equipped with atomic clocks that are synchronized with each other and with ground control stations, ensuring precise timekeeping.

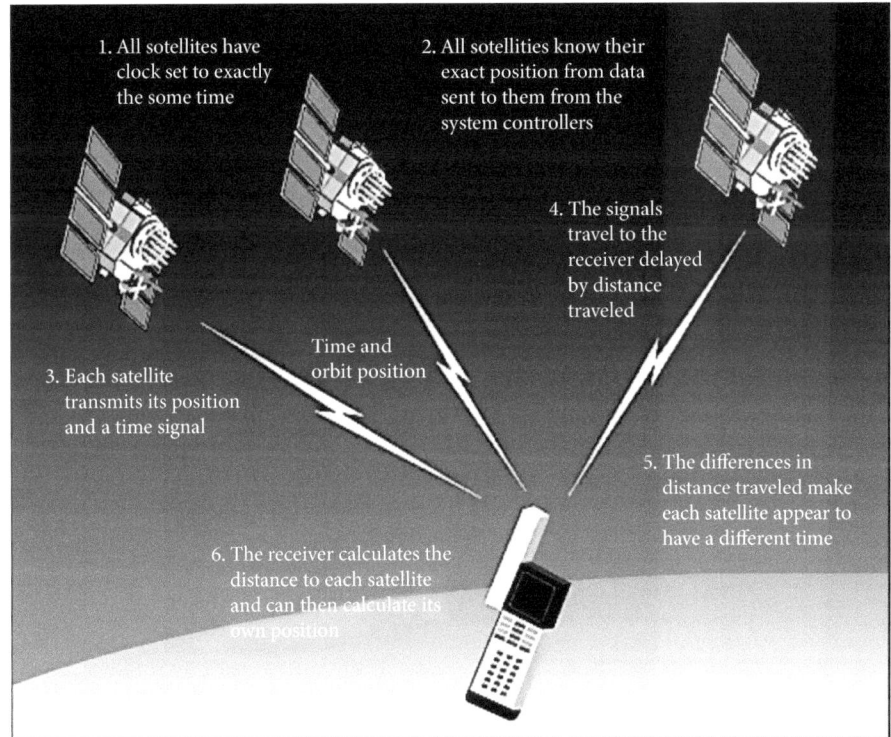

Figure 6.1 How satellite technology works.

Opaluwa, Y. D., Vincent C. Okorocha, Isaac C. Abazu, Joseph O. Odumosu, and G. O. Ajayi. "The Effect of GPS Satellite Geometry on the Precision of DGPS Positioning in Minna, Nigeria." *Jurnal Teknologi* (Sciences & Engineering) 77, no. 12 (2015). https://journals.utm.my/jurnalteknologi/article/view/6318

2. **Satellite Position Calculation**—Ground control stations continuously monitor and update the satellites' positions in space, allowing each satellite to determine and broadcast its exact location.
3. **Signal Transmission**—Each satellite transmits a time-stamped signal along with its position data.
4. **Signal Travel Time and Delay**—These signals travel at the speed of light, and their arrival time at the receiver depends on the satellite's distance: the farther the satellite, the longer the delay.
5. **Distance Calculation**—The receiver determines its distance from each satellite by measuring the signal's travel time and multiplying it by the speed of light.
6. **Position Determination**—By obtaining signals from four or more satellites, the receiver uses trilateration to calculate its precise three-dimensional position (latitude, longitude, and altitude) while using the fourth satellite to correct discrepancies in its internal clock.

This process takes place almost instantly, powered by advanced algorithms embedded in the receiver's processor. For smartphone users, this intricate calculation remains entirely behind the scenes—we simply see a blue dot appear on a map or receive accurate turn-by-turn directions. Yet, behind that blue dot lies a sophisticated system that transforms signals from orbiting satellites into precise location data.

7

Science by Satellite

Early Steps

Many of the initial uses of satellite technology for scientific purposes relied simply on its ability to locate and track objects on the ground with great precision. But before we get to those uses, we need to discuss one potentially fatal flaw in global positioning by satellite.

Relativity

In 1905, his *annus mirabilis*, or year of miracles, Albert Einstein postulated that time slows down for objects moving at speeds approaching that of light—a concept that became known as special relativity. Initially, many physicists dismissed the idea as counterintuitive, yet it has since been confirmed by numerous experiments. One of the most striking of these experiments involved flying atomic clocks around the world aboard commercial aircraft: the speedy airborne clocks were found to lose time relative to identical clocks that remained on the ground, just as Einstein had predicted.

A particularly vivid confirmation of relativity comes from the operation of the Global Positioning System. GPS satellites orbit Earth at an altitude of approximately 20,200 km and travel at speeds near 3.87 km/s—about 0.0013 percent the speed of light. Though this fraction is small, it is enough for special relativity to predict that satellite clocks should lose about 7 μs/day relative to clocks on Earth.

However, Einstein's general relativity adds an opposite effect. Because GPS satellites orbit well above Earth's surface, they experience a weaker gravitational field. According to general relativity, this causes their clocks to run faster—specifically, about 45 μs/day faster than identical clocks on Earth.

When both effects are combined, a GPS satellite clock runs approximately 38 μs/day faster than a clock on the ground. While this difference may seem trivial, failing to correct for it would lead to errors in GPS position calculations accumulating at a rate of about 10 km per day. The accuracy of GPS

Sentinels in the Sky. James Lawrence Powell, Oxford University Press. © James Lawrence Powell (2026).
DOI: 10.1093/9780197842850.003.0007

depends critically on incorporating these relativistic corrections into the system's software—making Einstein's theories an everyday necessity for modern navigation satellites.

Glaciers

Mountain and valley glaciers, distinct from the massive continental ice caps covering Greenland and Antarctica, are key indicators of climate change. When temperatures rise, these smaller rivers of ice shrink through melting, evaporation, and sublimation (the process by which a solid transitions directly into a gas, bypassing the liquid phase). As glaciers lose ice mass, their "snouts" retreat, while lower temperatures cause them to advance. The meltwater from shrinking glaciers flows into the oceans, contributing to rising sea levels, which are further amplified by the thermal expansion of warmer water.

Historically, glaciologists monitored changes by employing calibrated stakes, akin to surveyor's rods, embedded in the ice. Lateral shifts in the stake positions indicated the glacier's advance or retreat, while variations in the height of the stake above the ice revealed whether it was gaining or losing ice mass. Today, this labor-intensive method has mostly been replaced by highly precise satellite receivers mounted on glacier surfaces. These receivers continuously transmit data, enabling scientists to observe the position of the glacier's terminus, track ice flow speeds, and calculate changes in ice mass with minimal physical intervention, all in real time. This satellite-based technology offers a more efficient and accurate means to study the dynamic behavior of remote glaciers in a warming world.

It is no exaggeration to say that satellite observations have revolutionized the science of glaciology. The melting of the Antarctic and Greenland ice caps has drawn the most attention and represents, indeed, the greatest threat to humanity from climate change. However, the melting of the countless small mountain and valley glaciers is also significant, not only because it offers an early warning of the regional effects of climate change but also because many people around the world rely on glacial meltwater for their water supply. One region particularly vulnerable to melting ice is the west coast of South America, in the "rain shadow" of the Andes. The vast Peruvian city of Lima is one of the driest capitals globally, receiving an average of less than 10 mm (0.4 in) of rain annually. This precipitation primarily falls during the winter months as drizzle or mist rather than as substantial rainfall. How can a city of over 11 million people sustain itself when it is, meteorologically speaking, a desert? Historically, the answer has lain in the meltwater from mountain glaciers of the high Andes peaks, whose vulnerability now makes them "a fragile contradiction."[1] Between 2000 and 2016, the area covered by glaciers in the Peruvian Andes decreased by nearly 30 percent, with some glaciers vanishing entirely.

On a global scale, satellite observations indicate that between 2000 and 2019, the world's glaciers lost approximately 270 billion metric tons of ice each year. Compounding the problem are the satellite observations showing that the annual loss of glacial ice is accelerating. One study found that the *rate* of global ice loss—the yearly increase in the amount of ice melting—has itself been rising by approximately 6 GT (billion metric tons) per year on average.[2] If humanity's inaction allows global warming to continue its current trajectory, many mountain glaciers will disappear, further exacerbating the water supply crisis that climate change is certain to cause.

Plate Tectonics

One of the first and most obvious uses of GNSS for scientific purposes was to measure the movement of tectonic plates in near real time. The theory of plate tectonics had arisen mainly from the study of ancient rock magnetism, which showed that over millions of years, the seafloors had spread outward from the midocean ridges, sometimes carrying continents above them, confirming the long-rejected theory of continental drift. The magnetic evidence showed where the plates had been in the past, not where they are heading today. But ultra-precise satellite location measurements have made that possible, with results shown in Figure 7.1.

At first, geologists viewed tectonic plates as rigid slabs. However, satellites revealed a different reality: faults like the San Andreas bend and twist under stress. Each of these faults has triggered significant earthquakes. Before satellite technology became available, geologists could only measure fault movement using labor-intensive and inconsistent methods, such as surveying the positions of markers placed on adjacent plates and inferring their movements. Alternatively, they sent laser beams to reflectors and measured the return travel time, which gave more precise results. High-resolution satellite technology ushered in a new era of research into the intricate details of movement along faults. The carrier waves discussed in Chapter 5 enable the technology to provide 3-D data, revealing both vertical and horizontal movement. As satellite technology advanced, it gained the capability to acquire data every second, redefining "real time" and unlocking new possibilities for monitoring ground movement near earthquake zones.

As one example, geologists employing conventional surveying methods discovered that the North American and Pacific plates along California's San

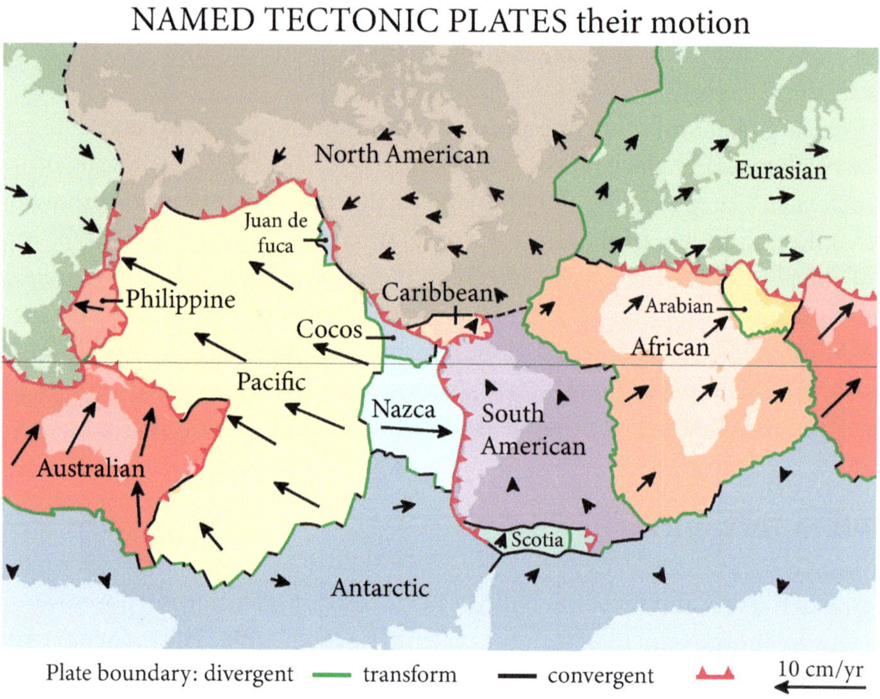

NAMED TECTONIC PLATES their motion

Figure 7.1 Plate direction and speed (length of arrow). Transform boundaries are where two plates slip past each other more or less horizontally.
Wikimedia Commons

Andreas Fault were slipping past each other at a rate of 47 mm per year.[3] However, satellite motion studies within each plate summed to a total of only 35 mm per year. Some process within the plates was absorbing the missing 12 mm, but what? A 1991 study using GPS, shortly after its full deployment, demonstrated that much of the difference was arising from rock deformation—folding and faulting—inside the North American Plate. This internal deformation absorbs some of the plate movement, leading to lower observed rates of slip between the plates. The satellite signal also revealed that the San Andreas Fault is not a single, simple fault line but a complex system of intertwined fault strands. Some slip may be occurring along these secondary faults, such as the Mojave Fault in the Basin and Range Province of southeastern California. Understanding this intraplate complexity is crucial for predicting where earthquakes might strike next in the state that awaits the "Big One."

Another example is the Hikurangi Subduction Zone offshore of New Zealand's North Island, where the Pacific and Australian plates converge (illustrated in Figure 7.2). Scientists studying the New Zealand fault system used continuous satellite signals, with networks of permanent receivers providing a steady flow of data. Geologists have long believed that earthquakes typically occur when the strain between two adjacent plates accumulates and is suddenly released. However, research in New Zealand and other fault zones using satellite technology has revealed a process known as "slow-slip," whereby the strain is released more gradually, over a timescale of hours to weeks rather than seconds to minutes. As shown, the New Zealand plate boundary features two oppositely directed subduction zones, where one plate slides beneath another. In the southern subduction zone, slow-slip events lasting up to a year occur approximately once every five years at depths of 30 to 50 km. In the northern zone, they happen much more quickly and at shallower depths. Both release similar amounts of energy but over quite different time frames. Such detailed knowledge would be impossible without continuous satellite signals. Researchers have discovered that beneath the southern section of the North Island, the subduction zone appears to be stuck, preventing movement and building up strain that could someday be released in a massive earthquake.

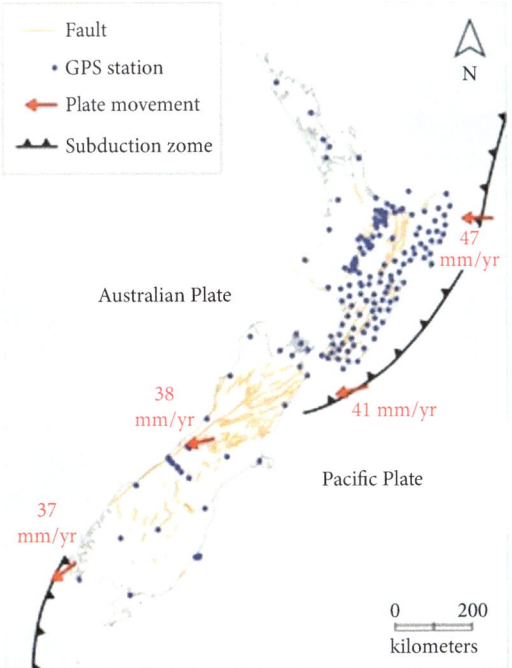

Figure 7.2 The double-facing subduction zones of New Zealand.

Earth Magazine

Earthquake Prediction

Satellite positioning can accurately determine the direction and speed of tectonic plate movements. However, it has not yet revealed *when* these movements will occur, though that time may be approaching. A critical development contributing to this possibility is the increased rapidity of satellite signal sampling. By the early twenty-first century, new receiver models could sample twenty to thirty times per second. This allowed geologists to study in detail how the ground shifted during the magnitude 7.9 Denali earthquake in Alaska in 2002.[4] Traditionally, ground motion during an earthquake has been measured with implanted seismometers, which detect and measure ground vibrations and movements caused by earthquakes and volcanic activity. However, those seismometers placed near the largest faults, like Denali, become "saturated" with so much data that they cannot keep up and so they effectively stop recording. Nevertheless, the high-rate sampling provided by the latest satellite receivers supplies the information that seismometers missed. This information has proven essential in studying even larger earthquakes. Satellite location revealed that the plates on either side of the 2010 magnitude 9.5 earthquake offshore of Chile, the largest ever recorded, moved up to 5 m horizontally and more than 1.7 m vertically. The 2011 Tohoku-Oki earthquake, a magnitude 9.1 event centered on the Pacific seafloor off the coast of Japan, ranks as the fourth most powerful earthquake ever recorded. It triggered the massive tsunami that devastated the Fukushima nuclear power plant. The quake caused extraordinary crustal displacements: it shifted the seafloor horizontally by 50 to 60 m and vertically by 5 to 10 m.

These studies have contributed to our understanding of giant earthquakes, but even more critical would be the ability to estimate the size of an earthquake virtually as it occurs, allowing for more accurate and timely warnings. One study of twelve large earthquakes using satellite technology found that within 10 seconds of the event, researchers could determine whether the quake would cause only minor damage, like the Denali quake, or be as devastating as the Tohoku-Oki.[5] (Earthquake magnitude scales by factors of ten, meaning that a magnitude 9.1 quake releases about 63 times more energy than a quake with a magnitude of 7.9.)

Rapid early estimates of magnitude could have saved lives in the December 2004 magnitude 9.3 Sumatra–Andaman earthquake, the third largest on record. A 1,000-km-long fault west of Sumatra ruptured, generating a tsunami that killed more than 250,000 people. Most casualties occurred on the nearby Sumatra mainland, where the wave reached a deadly height of 30 m.

Seismologists have long recognized that their traditional methods under-estimate the actual size of any earthquake larger than about magnitude 8.5. Seismometer saturation is not the only reason for this underestimation; the largest quakes often feature complex fault systems that a typical model of a single point source cannot adequately represent. Additionally, these quakes release a significant amount of energy at lower frequencies, which differently tuned seismometers may not capture fully. Moreover, the surface waves from an undersea earthquake may not be detected until they arrive at the near-est seismometers, which can be thousands of kilometers away. Therefore, by the time the ground waves reach those instruments, they have diminished in intensity. The initial seismological estimate for the 2004 Sumatra earth-quake suggested a major but not catastrophic event, with a magnitude of approximately 8.8 to 9.0. However, subsequent analyses using the magnitude scale revised the magnitude to between 9.1 and 9.3. Given the logarithmic nature of the magnitude scale, this increase corresponded to an earthquake that released approximately two to six times more energy than originally estimated. The final magnitude placed the event among the most powerful earthquakes ever recorded.

The underestimation of the Sumatra quake sparked renewed interest in using satellite technology to provide early warnings of an earthquake's prob-able magnitude. A 2005 study found that by measuring ground displacement with publicly available data from thirty-eight existing GNSS sites within 7,500 km of the epicenter, the Sumatra earthquake's actual magnitude and tsunami potential could have been determined within 15 minutes.[6] Had satellites been in place, the danger would have been apparent sooner and many lives could have been saved. This finding was particularly striking because few receiving stations were situated within 1,000 km of ground zero. Had there been addi-tional sites closer to the quake's epicenter, the warning time could have been reduced to as little as 10 minutes.

This example highlights the immediate "postseismic" period following an earthquake. However, earthquake science is approaching a new frontier: the potential to forecast earthquakes *before* they occur. One promising approach involves placing satellite receiving stations along major fault zones where the opposing sides appear locked, showing little or no seismic activity. By using today's highly precise satellite-based geodetic techniques, scientists can detect the slow, silent buildup of strain between these fault blocks—strain that may eventually be released in a major earthquake.

To be effective, natural hazard prediction methods must share two key characteristics: (1) the predicted event must be imminent, and (2) the pre-dictions must materialize nearly every time. Imagine geologists publicly predicting a quake of magnitude 8 or higher within the next 24 hours along the San Andreas Fault near San Francisco. This would trigger a massive evacuation of the city, which itself could result in significant loss of life, similar to the evacuation of the area around the Fukushima Daiichi reactor, which is estimated to have caused more deaths from disruption and stress than from the direct effects of radiation exposure. Now, suppose the predicted Bay Area "Big One" fails to occur. Governments, the public, and the media would be tempted to disregard the next warning, perhaps comparing the seismologists to Chicken Little, who mistook an acorn falling on his head for the sky falling. Scientists making predictions must err well on the side of caution and issue a warning only when they are nearly certain. However, so far, earthquake science has not provided that level of certainty.

The high sampling rate of the latest GNSS receivers has enabled the collection of vast quantities of precise, real-time location data. For example, analysis of data from 3,000 GNSS stations situated near ninety global earthquake epicenters—each with a magnitude above 7—revealed a striking pattern: on average, about two hours before rupture, the stations began to exhibit exponentially accelerating horizontal movement. This subtle, yet consistent, signal may offer a critical clue in the quest to detect earthquakes before they occur.[7] This brief period represents a life-saving improvement over the automatic earthquake early warning systems that currently provide only seconds to minutes after the quake has occurred. At least people could seek shelter, and as one author writes, people can be warned that "it is time to let go of sharp utensils and get ready to "Drop, Cover, and Hold On."[8]

Scientists have also begun to explore using satellite data to track the progress of tsunami as they speed across the seas, where in the deep ocean they can reach speeds of 500 to 800 km/h. Although this may seem incredibly fast, the oceans are so vast that it can still take a tsunami many hours to cross the open ocean, providing ample time to warn coastal communities.

Large tsunami displace such enormous volumes of water that as they move across the ocean, their interaction with the atmosphere above alters the total electron content (TEC) between the ionosphere and the satellite receiver. By tracking these TEC anomalies, scientists can, in theory, detect the passage of the tsunami-induced waves and, almost immediately after the submarine rupture that produced it, recognize that a tsunami has occurred. The magnitude 7.8 Haida earthquake in 2012 and resulting tsunami off Prince Rupert, Canada, provided a test case. TEC variations in the ionosphere measured at fifty-six stations in Hawaii, over 4,000 km from the epicenter, allowed scientists to determine the speed and direction of the atmospheric waves, showing that they could have been used as a tsunami predictor. Applying this method broadly would require a large increase in the number of GNSS receiver stations, but given what is at stake, the expense would be worth it.[9]

Volcanic Eruptions

Volcanic observatories increasingly rely on satellite technology to monitor subtle changes in the ground surface surrounding active volcanoes. These surface deformations reflect shifts within the underlying magma chamber and can offer critical early warnings of a potential eruption. During the 2018 eruption of Kīlauea on Hawaii's Big Island, such satellite-based observations were instrumental in guiding evacuation efforts by identifying regions at greatest risk. After the eruption, researchers determined that approximately 0.8 km³ of magma had been released—equivalent to about 320,000 Olympic-sized swimming pools. This estimate closely matched posteruption field measurements, confirming that satellite observations of ground deformation can yield accurate and actionable information.[10]

Volcanic ash plumes can be deadly—Pompeii proved that. Even if they do not result in loss of life, these plumes can have significant effects, as evidenced by the 2010 eruption of the Eyjafjallajökull volcano in Iceland. Its plume disrupted air travel across Western and Northern Europe for about six days. Over 100,000 flights had to be canceled, affecting approximately 10 million travelers. This led to an estimated loss of $1.7 billion for the airline industry, along with additional costs for other businesses that rely on "just-in-time" deliveries. For the Icelanders, the eruption caused an immediate decline in foreign visitors, but at least some of this loss was offset by subsequent "volcano tourism."

Like earthquakes, scientists need to be able to quickly assess the size and direction of volcanic ash plumes. Traditionally, this assessment has been done using satellite photography, but that method is effective only during the daytime and can be worsened by weather conditions. But as these plumes rise into the atmosphere, they disrupt signals traveling between satellites and ground-based receivers. By analyzing changes in the delay and the strength of the satellite signal, scientists can infer the presence, density, and movement of a volcanic ash plume.

In a 2013 article, Kristine Larson of the University of Colorado and her colleagues used satellite data to examine the volume of ash emitted by eruptions in Alaska and the speed at which the resulting plume traveled.[11] Her findings aligned with the estimates from radar and other sources, confirming the method's effectiveness in detecting volcanic ash plumes and their trajectories.

Additionally, scientists are investigating the potential of crowdsourced smartphones equipped with built-in accelerometers to measure ground motion and monitor volcanic activity.

Reflectometry

Satellite receivers not only record the incoming signals from satellites overhead, as they are supposed to, but they also inevitably capture signals that have been reflected off the ground, nearby buildings, and other surfaces. For years, scientists dismissed these reflected signals as mere noise. They ultimately realized they could be put to good use. In 1993, Manuel-Martín Neira, a researcher with the European Space Agency, proposed that direct and reflected satellite signals could be used in tandem to provide new information that was otherwise unavailable.[12] His proposed method of "passive reflectometry" has led to a surprising number of new and important applications (see Figure 7.3), especially in remote areas. When a satellite transmits a signal, part of it travels directly to a stationary receiver on the ground, while another portion reflects off the snow surface before bouncing back to the same receiver. Because the reflected signal follows a longer path—down to the snow surface and then back up—it arrives slightly later than the direct signal. By precisely measuring this time delay, scientists can calculate the additional distance traveled by the reflected wave. Since the geometry of the satellite-receiver pair is known, this difference reveals the height of the reflecting surface—allowing the snow depth to be estimated relative to the known ground level beneath it. The greater the delay, the deeper the snow.

This method effectively monitors snow depth across vast and inaccessible areas, such as mountain ranges and polar regions, and it is now widely used in climate studies and hydrology. By integrating snow depth data with other observations, researchers can track changes in snowpack, which are vital for managing water resources, predicting floods, and assessing the vegetation-related impacts of global warming, such as drought stress and land-cover change.

The reflected signal also provides information about the density and moisture content of the snow, which facilitates calculation of the "snow water equivalent": the amount of water that would be produced were a snowpack to melt completely. This enables the prediction of future runoff into streams and reservoirs. Larson tested this technique on areas of approximately 1,000 m^2 and found that it could measure snow depth within an accuracy of about 0.4 m.[13]

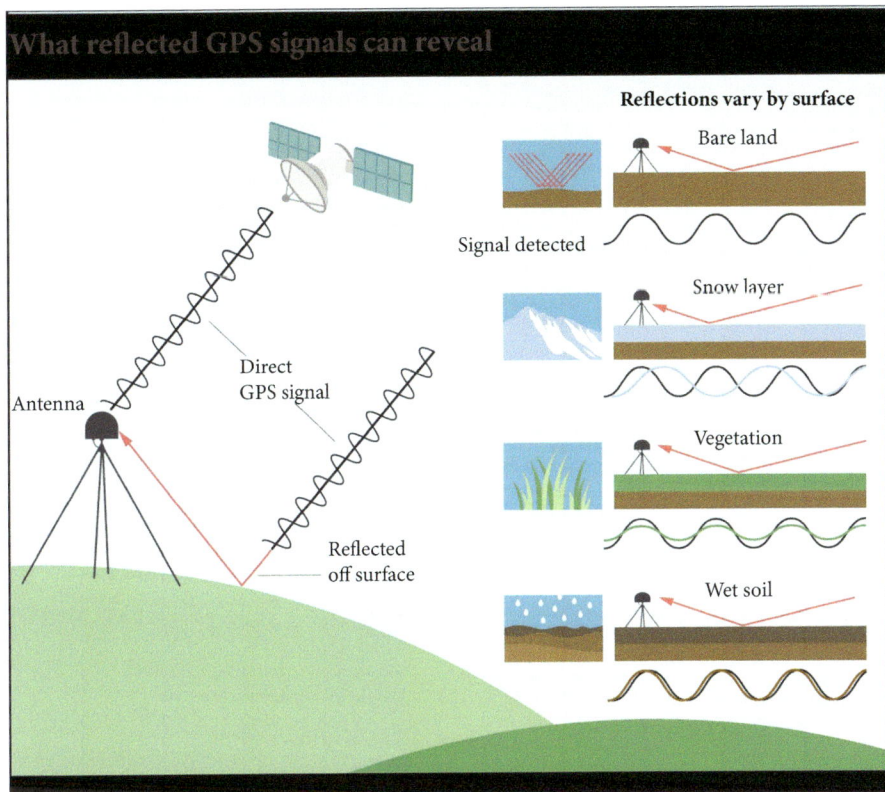

What reflected GPS signals can reveal

Reflections vary by surface

Bare land

Signal detected

Snow layer

Antenna

Direct
GPS signal

Vegetation

Reflected
off surface

Wet soil

Figure 7.3 Schematic showing the use of the reflected satellite signal to detect bare land and to measure snow depth, vegetation cover, and soil moisture.
Knowable Magazine; UNAVCO

Water vapor in the atmosphere slows down satellite signals, causing slight delays as they pass through. By analyzing these delays, scientists can estimate the amount of "precipitable water"—the total atmospheric moisture available to fall as rain or snow. In Southern California, a denser network of GNSS receiver stations has greatly improved these measurements, especially during the summer months when intense storms are more likely. This enhanced capability allows for more accurate forecasting and timely warnings of flash floods, debris flows, and road closures, helping to mitigate the impacts of sudden, weather-driven hazards.

In another inventive application, Larson and colleagues examined how satellite signals reflected off the sea surface in an area of Alaska known for having some of the largest tides in the United States. This and other studies demonstrated that satellite technology can measure tidal variations as accurately as traditional gauges. One study found that over ten years, daily tidal measurements from the two methods agreed within 2 cm.[14] Utilizing satellite signals has the advantage over use of conventional tidal gauges in that it is always active and does not require on-site measurements.

As a reminder, in this chapter, we have examined some direct applications of satellite technology—those that do not require additional instruments. From here on, we will explore the wide range of discoveries made using instruments carried aboard satellites, starting with cameras.

PART III

STORMY WEATHER

8

Harry Wexler's Dream

Mark Twain famously quipped that everyone talks about the weather, but no one does anything about it. That comes as no surprise, since for most of human history, even if people had been able to influence the weather by some miracle, they lacked the information necessary to know what to do. Before the invention of the telegraph, weather knowledge was limited to local observations or secondhand accounts shared by travelers, who were likely to report that the weather in the next county over was about the same as in their county. News of extreme weather events in more distant locations could take weeks to arrive, if ever it did. This lack of timely and reliable information rendered accurate weather forecasting virtually impossible.

Today, meteorologists can predict weather patterns with impressive accuracy up to two weeks in advance, a time frame unimaginable in Twain's era. These warnings not only help us prepare for typical conditions but also save lives by forecasting dangerous weather events like hurricanes, tornadoes, and blizzards long before they strike.

The Telegraph: The Pioneering Weather Instrument

In 1844, painter and inventor Samuel Morse installed a telegraph line between Baltimore and Washington, D.C. To transmit messages, he used the simple dot-dash code system that he invented and that carries his name.[1] His inaugural message, "What hath God wrought?" from Numbers 23:23, may have seemed overblown; however, it foreshadowed the emergence of the telegraph as not just another technological invention, but one that would transform many aspects of human life. Telegraphy was one of the first ways in which human ingenuity overcame the limitations imposed by distance, later followed by radio, the airplane, television, satellites, and the internet.

Within four years, 2,100 miles of telegraph lines crisscrossed America—rapid by any standard and revolutionary in the 1840s. During the subsequent Civil War, the telegraph would prove invaluable to both sides, providing updates on distant battles and informing commanders about enemy troop concentrations.

Sentinels in the Sky. James Lawrence Powell, Oxford University Press. © James Lawrence Powell (2026). DOI: 10.1093/9780197842850.003.0008

Joseph Henry, the first director of the Smithsonian Institution, immediately recognized the potential of the telegraph for gathering weather information. In exchange for weather data, he offered to provide equipment to the many telegraph offices that were springing up. In 1855, the Smithsonian inaugurated a new headquarters building on the National Mall. The lobby prominently featured a large weather map displaying real-time weather conditions from across the country. To standardize data collection and reporting, Henry devised comprehensive instructions for weather observers. His guidelines detailed precise methods for observing, recording, and transmitting data on temperature, barometric pressure, wind direction, and other atmospheric phenomena, ensuring consistency and accuracy in meteorological reports.

As the data flowed in, they were plotted on the Smithsonian's lobby map, revealing weather patterns and the movement of weather systems across the country, a necessary first step toward forecasting. As the Smithsonian's directors reported in 1858, "This map is not only of interest to visitors, showcasing the kind of weather their distant friends are experiencing, but is also important for determining at a glance the probable changes that may soon be expected."[2]

By 1860, some 500 U.S. weather stations were participating in the effort.[3] Then came the Civil War, during which Southern stations, unsurprisingly, ceased sending data. However, the groundwork had been laid for establishing the U.S. National Weather Service (NWS) in 1870. The purview of the NWS included collecting observations from various locations, analyzing data, and issuing forecasts to protect life and property, particularly in maritime and agricultural activities. One crucial achievement of the early weather service was its role in significantly reducing the loss of life at sea through timely storm warnings. Henry's pioneering work serves as a prime example of how a modest investment in science and technology can yield disproportionate benefits. As Mark Twain is also quoted as saying, "Science has yielded such a great harvest of truth to the world that no expenditure in time or money has been too extravagant."

Early Weather Forecasting

In the early twentieth century, a group of Norwegian scientists, led by physicist and meteorologist Vilhelm Bjerknes, established what became known as the Bergen school of meteorology, named after their home city. Their focus was on the mathematics and theory of large-scale weather patterns. They viewed weather primarily as the result of the movement and interaction of warm and cold air masses separated by "fronts." They found that storms move predictably along these boundaries, making weather forecasting possible—at least in theory. Bjerknes was the first to formulate a set of mathematical equations describing the flow of the atmosphere. However, as he acknowledged, solving his equations "far exceed[ed] the means of today's mathematical analysis."[4] In contemporary weather forecasting and climate modeling, supercomputers now solve these difficult equations. Another challenge for meteorologists in Bjerknes's time was that there were too few reporting stations to provide the range and continuity of data necessary for building adequate prediction models. By the 1950s, however, weather stations worldwide were collecting surface readings four times a day.

At first, meteorologists used kites and balloons to gather atmospheric data. By the twentieth century, a breakthrough came with the radiosonde, a small, battery-powered instrument carried aloft by a weather balloon to measure temperature, humidity, pressure, and wind, transmitting real-time data back to the ground via radio signals. These devices revolutionized atmospheric science, providing continuous, high-altitude observations essential for weather prediction.

Today, thousands of radiosondes are launched daily by meteorological agencies worldwide, following a schedule set by the World Meteorological Organization (WMO). They return essential data on the vertical structure and dynamics of the atmosphere. Despite the rise of weather satellites, radiosondes remain indispensable, serving as critical tools for calibrating satellite sensors, validating remote sensing data, and providing direct atmospheric measurements that satellites can only infer.

First Views from Space

To understand and forecast the weather, it is essential to observe from an altitude high enough to encompass an area roughly the size of weather patterns. Before the post-World War II period, the highest photograph of the atmosphere ever taken came in 1935 from Explorer II, a helium-filled balloon mission funded by the National Geographic Society and the Army Air Corps (not to be confused with the later Explorer satellite missions; see Figure 8.1). Using an aluminum gondola to minimize weight and carry its two pilots, Explorer II set a new altitude record of 22,066 m. Upon their return to Earth, the two pilots were feted as national heroes, again foreshadowing the acclaim that would greet returning astronauts a few decades later.

Figure 8.1 Explorer II gondola in the National Air and Space Museum. Launched on November 11, 1935, from near Rapid City, South Dakota, it reached a record altitude of 22,066 m.
Wikipedia Commons

The 1940s ushered in a new era of exploration, as rockets soared to unprecedented heights, delivering the first breathtaking images of Earth from above. Unsurprisingly, the first rocket mission to achieve this was led by von Braun's team at White Sands.[5] On October 24, 1946, they launched a captured A-4 with a movie camera in its nose cone. It ascended to 105 km before slowing to a halt, then crashing into the New Mexico desert. The steel casing that protected the film canister was recovered from the wreckage to provide the photo shown in Figure 8.2.

Figure 8.2 First photo of Earth from space, from a captured V-2 rocket launched by von Braun's team at White Sands.
Smithsonian Magazine, Applied Physics Laboratory

In the early days of rocketry, cameras were mounted on launch vehicles primarily to monitor their orientation during key phases of flight. To ensure clear images of the ground for navigation and tracking, these launches were typically scheduled on cloudless days. As a result, the earliest photographs from rockets showed only the terrain below, giving little or no indication of weather patterns or atmospheric conditions. This approach began to change on October 5, 1954, with the launch of an Aerobee rocket from White Sands Missile Range. Developed by the Applied Physics Laboratory of Johns Hopkins University, the Aerobee was designed to carry scientific instruments into the upper atmosphere and near-space, marking a transition from imagery for engineering or navigational purposes to scientific observations of the Earth's atmosphere and weather from above. It consisted of two stages: a solid-propellant booster and a liquid-fueled upper stage. This combination allowed the Aerobee to reach an altitude of 230 km, well above the 100 km used to define the boundary of space: roughly the altitude a winged craft can reach before the atmosphere becomes too thin to provide lift. The Aerobee

Figure 8.3 A photomosaic taken by the Aerobee rocket on October 5, 1954 showing a band south and east of the launch site at White Sands, New Mexico, and a tropical cyclone.
Wikimedia Commons

family of rockets laid the foundation for America's space program and provided a wealth of data on cosmic rays, solar radiation, and the atmosphere that could not have been obtained in any other way at the time. The Aerobee carried two 16-mm movie cameras in its nose cone, mounted perpendicular to the length of the rocket. In one instance, at about 160 km altitude, the craft rolled, allowing the cameras to point downward and photograph bands of Earth's surface, shown in Figure 8.3. The nose cone was then released and parachuted back to Earth, where the film was recovered.

The Aerobee photographs, along with others from early rockets, hinted at the valuable insights that could be gained from observing weather patterns from high altitudes. However, rockets alone could not provide a comprehensive account of weather phenomena. Their unpredictable movements hindered the onboard camera from focusing on specific locations or even aiming upward instead of downward. Furthermore, rockets were costly. Most importantly, to effectively understand and predict the ever-changing weather conditions, photographs needed to be delivered promptly and multiple times a day.[6] It became evident that a continuous stream of data, capable of being transmitted to meteorologists, was essential. This realization would soon pave the way for the use of satellites to predict weather.

In 1951, the RAND (Research and Development) Corporation, an early and influential "think tank," published a report titled "Inquiry into the Feasibility of Weather Reconnaissance from a Satellite Vehicle."[7] Released at the dawn of the Cold War, the report framed weather satellites as a military necessity. Monitoring enemy skies, after all, was just as important as predicting storms. The opening lines read: "It is assumed that, in the event of armed conflict, aerial weather reconnaissance over enemy territory, similar to that obtained in World War II, will be extremely difficult if not impossible. An alternative method of obtaining this information, however, is thought to lie in the use of the proposed satellite vehicle."

Harry Wexler: The Visionary

Meteorologist Harry Wexler saw that satellites could transform how humanity perceived and predicted the weather. He graduated from Harvard and earned a PhD in meteorology from MIT in 1939, served as a weather officer during the war, and later returned to the U.S. Weather Bureau, where he worked for the remainder of his career. In 1954, Wexler delivered a talk on the advantages of satellites for meteorology. Encouraged by a letter from Arthur C. Clarke,[8] who was undoubtedly pleased to see his earlier visionary ideas catching on, Wexler published the talk under the title "Observing the Weather from a Satellite Vehicle."[9]

Wexler summarized the various types of geophysical and solar data he believed satellites could provide:

(a) Surface and atmospheric temperature—Measuring the temperature of Earth's surface and estimating the average temperature of the overlying atmosphere by observing the infrared spectrum.

(b) Precipitation—Detecting rainfall, potentially through the use of radar.

(c) Thunderstorm activity—Locating lightning either visually (at night) or electronically (during the day).

(d) Solar radiation—Measuring solar output to investigate possible correlations with weather patterns and determine whether unusual weather events are influenced by solar activity.

(e) Albedo—Monitoring day-to-day changes in Earth's reflectivity to solar radiation, which could yield valuable climate information.

(f) Meteoric dust—Detecting particles that, according to a prevailing theory at the time, might act as cloud-seeding agents and contribute to increased rainfall.

"In summary," Wexler concluded, "it can be stated without question that a satellite vehicle, moving about Earth at the proper height and manner, would be of inestimable value as a weather patrol for short-range forecasting and as a collector of basic research information for solar and geophysical studies, including long-term weather changes and climatic variations."[10]

In order to convince other scientists and government officials of the value of weather satellites, Wexler did something unusual for a scientist of his or any other day: he commissioned a colored drawing. As shown in Figure 8.4, the drawing depicted Earth as it might be viewed from 5,450 km above Amarillo, Texas, at noon on June 21, 1954.[11] The drawing was lost for years until scholar James Rodger Fleming discovered it among the Wexler papers in the Library of Congress. It illustrated a variety of weather features, including a small hurricane north of Puerto Rico, a front between the northeast trade winds of

Figure 8.4 Wexler's Dream.
NOAA's National Environmental Satellite, Data, and Information Service (NESDIS)

the Northern Hemisphere and the southeast trades of the Southern Hemisphere, and a breeding ground for severe windstorms and tornadoes known as a "line-squall."

By 1957, Wexler's dream was no longer mere speculation. Sputnik demonstrated that satellites could reach orbit, and it was only a matter of time until the first U.S. weather satellite was launched. High-altitude photography had already commenced with the Aerobee and other rockets, and soon there would be no technological barriers to capturing even higher-altitude images of Earth's weather patterns.

9
Weather Satellites

Explorer VII

Sputnik caught America off guard, but within months, U.S. scientists were ready to launch satellites of their own—not only to rival the Soviets but also to study Earth's weather from space.[1] Explorer VII, launched in 1959, proved that satellites could track vast weather systems in real time—an impossibility just two years earlier. It included various scientific experiments, as shown in Figure 9.1, such as a set of temperature sensors to measure the radiation emitted by the Sun and the amounts absorbed, reflected, and re-radiated—just the type of data Harry Wexler had envisioned collecting once satellite launches became a reality. Explorer VII was far more sophisticated than Sputnik I, illustrating how far satellite technology had advanced in only two years.

One noticeable difference was that, unlike the spherical Sputnik, Explorer VII featured arms and various protrusions, appearing as unstreamlined as one could imagine. At first glance, this feature seems to contradict common sense: shouldn't these speedy objects be streamlined, like a fast-moving car, train, or plane? Then, we realize that for an object in the vacuum of space, there is no friction regardless of its shape. In that environment, streamlining offers no advantage.

Unless its orbit can be stabilized, a satellite released at high altitude will tumble unpredictably, capturing images in random directions instead of focusing on specific weather systems or regions. To prevent this outcome, early satellites, including Sputnik and Explorer VII, relied on spin stabilization. The principle can be understood by considering a spinning top: when first set in motion, the top wobbles briefly before settling into a steady, upright spin. It remains stable in this position due to its angular momentum, a conserved quantity described by Newton's laws, meaning the top will keep spinning unless acted upon by an external force. However, an earth-bound top eventually succumbs to such forces—gravity and friction between its stem and the surface, for example—causing it to lose energy, wobble, and soon topple over. In contrast, a satellite in space, once spun up by small thruster engines, can remain stable almost indefinitely. The disadvantage is

Sentinels in the Sky. James Lawrence Powell, Oxford University Press. © James Lawrence Powell (2026).
DOI: 10.1093/9780197842850.003.0009

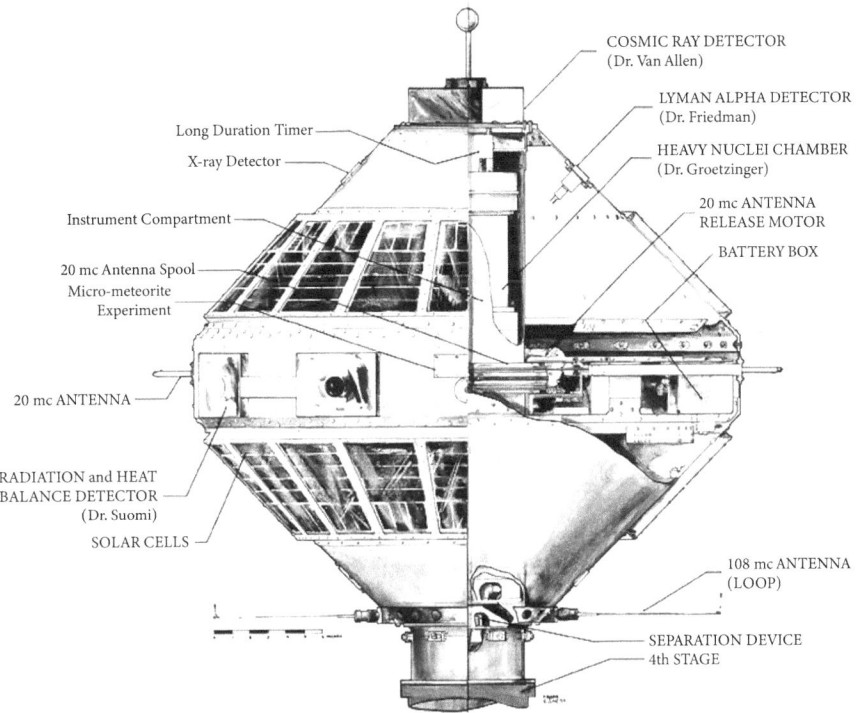

COSMIC RAY DETECTOR
(Dr. Van Allen)

LYMAN ALPHA DETECTOR
(Dr. Friedman)

Long Duration Timer

X-ray Detector

HEAVY NUCLEI CHAMBER
(Dr. Groetzinger)

20 mc ANTENNA
RELEASE MOTOR

Instrument Compartment

BATTERY BOX

20 mc Antenna Spool

Micro-meteorite
Experiment

20 mc ANTENNA

RADIATION and HEAT
BALANCE DETECTOR
(Dr. Suomi)

SOLAR CELLS

108 mc ANTENNA
(LOOP)

SEPARATION DEVICE
4th STAGE

Figure 9.1 Schematic of Explorer VII showing the many different instruments it carried.
Wikimedia Commons

that a spin-stabilized satellite cannot quickly change its orientation to point at specific targets. Later methods of maintaining stability did not have that drawback, but for the early, less ambitious satellites, spin stabilization was a good first step.

TIROS: The Next Generation

The next phase of America's space program was TIROS (Television Infrared Observation Satellite). The program deployed satellites in Low Earth Orbit (LEO), which ranges from approximately 160 km to 2,000 km in altitude. At these heights, satellites complete an orbit in 90 to 120 minutes, with their exact speed depending on altitude. As Earth rotates beneath them, different regions come into view with each circuit. This design allows polar-orbiting TIROS satellites to revisit the same area at the same time each day, making them ideal for tracking changes in weather systems over time. In April 1959, the newly formed National Aeronautics and Space Administration (NASA) took over development of the TIROS program from RAND. After extensive ground testing, TIROS-1, illustrated in Figure 9.2, was launched from Cape Canaveral, Florida, on April 1, 1960, marking the start of the modern era in satellite-based weather prediction.

As shown in Figure 9.2, the inaugural TIROS satellite was an eighteen-sided cylinder, its surface covered in solar panels except for its base. To the engineer, it was a thing of beauty. At a time when both men and women often wore hats, one wag quipped that TIROS resembled a familiar hat box. The lower section housed transmission antennas, while a single reception antenna was mounted on top. A unique de-spin mechanism reduced the spin rate when

Figure 9.2 TIROS-1. The elongated metal cylinders around the base are the thruster engines used to maintain spin. The large cylinder extending down from the base at the lower left is one of the two cameras carried by TIROS-1.
National Air and Space Museum

necessary for photography. A pair of tethers, each attached to a weight, were coiled around the satellite. Once in orbit, the weights would be released—like anchors dropped at sea—causing the tethers to unwind and gradually slow the satellite's spin. As the tethers were reeled back in, the spin would accelerate again, with the system conserving angular momentum throughout.[2]

TIROS-1 was equipped with two cameras: one wide-angle and one narrow. Within hours of launch, it transmitted the first-ever weather image from space, as shown in Figure 9.3. The images clearly displayed the New England coast and Canada's Maritime Provinces, extending north to the St. Lawrence River. These cameras used miniature vidicon tubes, each capable of capturing a 500-line image in seconds. The imagery could be either immediately telemetered to a ground station or recorded on magnetic tape for later transmission when the satellite passed over a station. Each camera had its own recording system, transmitter, and electronic circuitry, which enabled autonomous operation. One camera featured a wide-angle lens for capturing expansive views, while the other offered more focused narrow-field imaging.

The photos were airlifted promptly to Washington, D.C., to be presented to President Eisenhower. "We've suddenly gone from rags to riches overnight," Harry Wexler said, witnessing his dream come true.

In 1961, the TIROS-3 satellite made history by photographing Hurricane Esther, shown in Figure 9.4, marking a significant advance in meteorological observation. Equipped with advanced imaging technology for its time, TIROS-3 tracked the storm's unusual trajectory, which included a rare

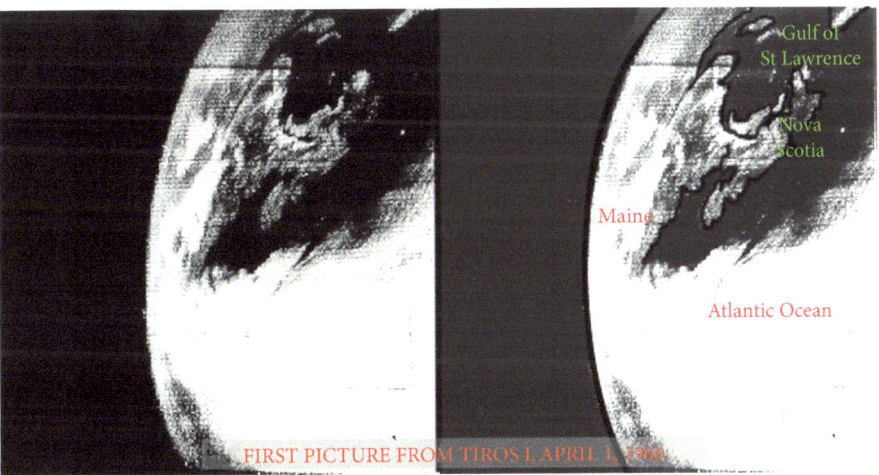

Figure 9.3 First photo from TIROS-1.
Wikimedia Commons

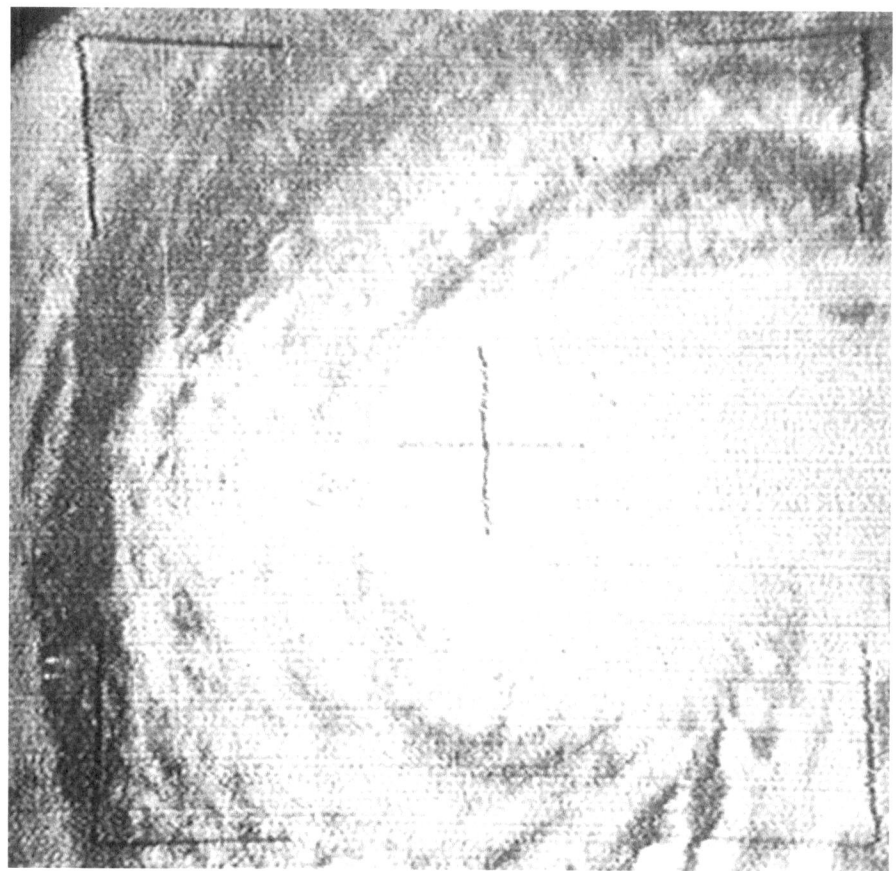

Figure 9.4 Hurricane Esther, the first hurricane to be detected from space, TIROS-3,
September 10, 1961.
Wikimedia Commons

loop-the-loop over the Atlantic Ocean. This unexpected path allowed scientists to observe the hurricane's behavior in unprecedented detail, providing valuable data on its evolution, intensity fluctuations, and interactions with the atmosphere.

1. The very first TIROS images demonstrated the advantages of weather satellites:
2. The satellites could be placed in a high enough orbit to reveal a swath of Earth roughly the size of weather patterns.
3. The images can be telemetered to meteorologists in real time, allowing up-to-date weather maps and forecasts.

4. The images clearly show clouds and other weather features, including tropical cyclones, before conventional methods could detect them.

President Eisenhower may have been prescient when he noted another feature that the TIROS satellites revealed: "The Earth doesn't look so big when you see that curvature." Many other observers would have a similar thought and would perhaps have paused to consider the uniqueness of our precious celestial home when they saw the famous "Blue Marble" photo of Earth in space taken on Apollo 17 on December 7, 1972.

Orbits

Let us pause here to better define the different types of satellite orbits shown in Figure 9.5. The top center frame illustrates a polar orbit, in which a satellite travels over, or nearly over, both the North and South Poles on each revolution. As Earth rotates beneath this orbital path, the satellite gradually scans different swaths of the planet's surface. This allows for global coverage over time, making polar orbits especially useful for monitoring weather systems, analyzing atmospheric conditions, and mapping Earth's surface with high precision.

A geostationary orbit is a special case of a geosynchronous orbit in which a satellite completes one revolution around Earth in exactly one sidereal day—23 hours, 56 minutes, and 4 seconds—returning to the same position relative to the stars. Positioned directly above the equator at zero degrees inclination and an altitude of 35,786 km, the satellite appears stationary over a fixed point on Earth's surface. This stability is maintained through periodic thruster adjustments that compensate for orbital drift. These geostationary satellites are vital for continuous, real-time monitoring of large areas, and they are used extensively for weather observation, environmental surveillance, and telecommunications, including television broadcasting, internet connectivity, and telephone networks. In addition, they play a crucial role in military and security operations, enabling the tracking of ships, aircraft, vehicles, and troops, as well as the monitoring of national borders and support for strategic missions.

The TIROS satellites were in *sun-synchronous or inclined orbits* (shown in the right frame of Figure 9.5), a type of near-polar orbit typically located at altitudes between 600 and 800 km. In these orbits, the satellite crosses any given point on Earth's surface at the same local solar time each day. This synchronization is achieved by giving the satellite a specific inclination

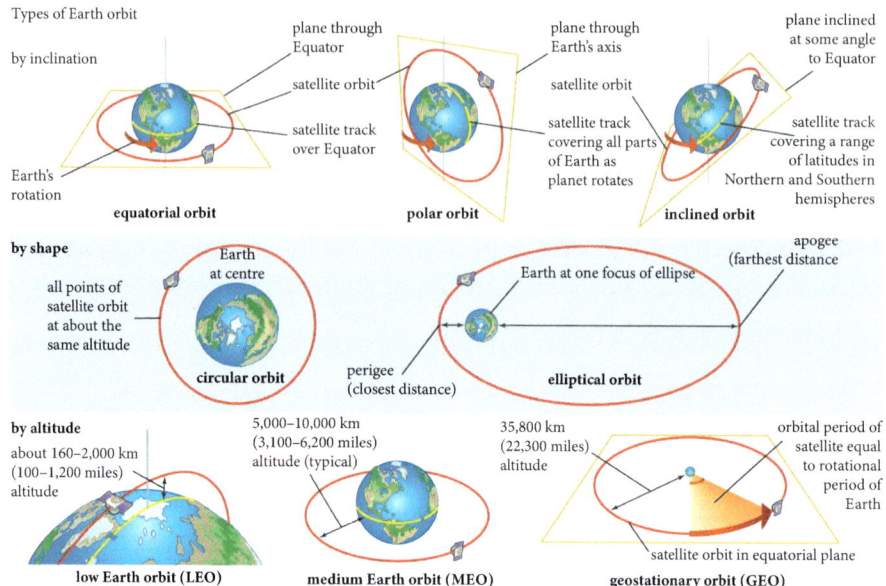

Types of Earth orbit

by inclination

Earth's rotation

plane through Equator

satellite orbit

satellite track over Equator

equatorial orbit

plane through Earth's axis

satellite orbit

satellite track covering all parts of Earth as planet rotates

polar orbit

plane inclined at some angle to Equator

satellite track covering a range of latitudes in Northern and Southern hemispheres

inclined orbit

by shape

all points of satellite orbit at about the same altitude

Earth at centre

circular orbit

perigee (closest distance)

Earth at one focus of ellipse

apogee (farthest distance

elliptical orbit

by altitude

about 160–2,000 km (100–1,200 miles) altitude

low Earth orbit (LEO)

5,000–10,000 km (3,100–6,200 miles) altitude (typical)

medium Earth orbit (MEO)

35,800 km (22,300 miles) altitude

satellite orbit in equatorial plane

geostationary orbit (GEO)

orbital period of satellite equal to rotational period of Earth

Figure 9.5 Satellite Orbits.
Encyclopedia Britannica

and altitude that causes the orbit to precess (or rotate) slightly each day to align with Earth's movement around the Sun, ideal for satellites engaged in environmental monitoring and reconnaissance.

NIMBUS and the Applications Technology Satellites

By the mid-1960s, though TIROS continued to provide reliable weather observations through television imagery, two new programs were introduced: NIMBUS (Latin for "cloud") and ATS (Applications Technology Satellites). The NIMBUS program consisted of seven satellites, the first of which was launched in August 1964 and the last in 1978. NIMBUS was more experimental than TIROS, carrying a variety of new and improved instruments along with a newly designed stabilization and control system. Its sun-synchronous orbit enabled NIMBUS to provide consistent, repeatable observations of Earth's surface under similar lighting conditions, gradually building global coverage over time.

The Earth receives energy from the Sun primarily in what is known as shortwave radiation in the visible and ultraviolet regions of light, as illustrated in Figure 10.1. The Earth's surface absorbs this solar energy, warms up, and subsequently re-emits the energy, but as longwave radiation (infrared radiation). Carbon dioxide in the atmosphere preferentially absorbs long-wave radiation and radiates some of it back down, blanketing and heating our planet. NIMBUS measured a critical climate variable: Earth's albedo, or reflectivity, which represents the balance between incoming solar radiation and the amount Earth absorbs and re-emits. The amount of solar radiation that Earth absorbs directly affects its temperature: more absorption leads to warming, while greater reflectivity causes cooling. As global warming progresses, changes in Earth's albedo—its reflectivity—reduce the amount of sunlight reflected back into space. This is especially true as ice melts and land use changes, both of which lower the planet's overall reflectivity. The result is a positive feedback loop: less reflection leads to more heat absorption, which accelerates warming, causing even further declines in albedo. Accurately understanding and continuously monitoring albedo trends is therefore crucial for predicting both the rate and geographic distribution of climate change.

A striking discovery made by NIMBUS 1 was an unmapped black spot that suddenly appeared off the southwest coast of Iceland. It turned out to be a newly formed volcanic island, given the name of Surtsey for Surtr, a fire giant in Norse mythology The satellite was able to track the island's growth in real time, offering unprecedented insight into volcanic processes on a midocean plate boundary.

One key advantage of NIMBUS, particularly compared to the TIROS series, was its ability to capture thermal images at night using a heat-sensing radiometer. In these nocturnal images, warm land and water appeared dark, while clouds were lighter in color, enhancing the ability to monitor atmospheric and surface conditions around the clock.

The next advance was NASA's Applications Technology Satellites (ATS), whose cameras photographed an entire hemisphere continuously. Five launches were planned, but only two were successful. ATS-1 transmitted weather information to ground weather stations via facsimile (FAX). These satellites proved especially important as they transmitted educational programs that could be received anywhere, including remote Alaskan villages.[3] These hamlets often had few medical services or doctors, and in at least two instances, ATS-1 helped to save lives. In one case, a medical aide received a doctor's instructions via ATS-1 to relieve a person suffering from an appendicitis attack. In another, a nurse was advised on how to treat a pregnant woman who was hemorrhaging. Both patients survived and were taken to hospitals for further treatment. ATS-3 launched in November 1967, carrying a camera capable of capturing color images. It took one of the first full-disk

Figure 9.6 First color photograph encompassing the whole Earth taken by the ATS-3 satellite on November 10, 1967.
Wikimedia Commons

color photographs of Earth, preceding by a year the famous Earthrise images from Apollo 8 in 1968 (see Figure 9.6.)

ATS-3 also broke new ground in communication, transmitting color television from the 1968 Mexico Olympics and the Apollo 7 mission. Its geostationary orbit enabled meteorologists to observe the formation and evolution of weather systems, even allowing them to create films of the movement of the patterns. These films revealed that just before tornadoes occur, thunderstorm clouds expand dramatically, providing a useful precursor. By the end of the 1960s, satellites were able to detect almost all tropical storms before traditional methods could do so.

Modern Weather Forecasting

By the early twenty-first century, enhanced computer power and a wealth of satellite data enabled forecasts ranging from three to seven days. These forecasts could accurately predict general weather patterns, such as the

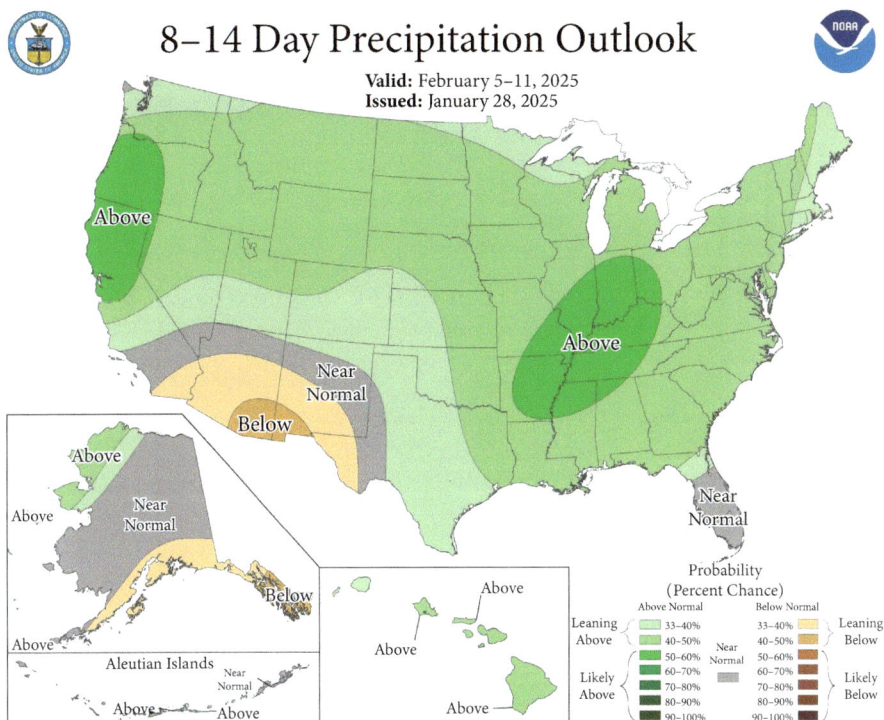

Figure 9.7 The 8–14 day precipitation forecast for the United States made on January 28, 2025.

National Weather Service

likelihood of rain, periods of cold or warm weather, and significant storm systems. Today's forecasts extend even further—reaching up to ten to fourteen days ahead—as illustrated in the weather map in Figure 9.7.

Fifty Years of Progress

Figures 9.8 and 9.9 show two images, one from 1965 and the other from fifty years later, illustrating the advances in satellite-based Earth observation and weather forecasting.

At the time that Figure 9.8 was taken, TIROS-IX was in a polar orbit, circling Earth every two hours. With each pass, its two cameras captured forty images of the sunlit side. One month after its launch, NASA scientists assembled 450 images to create the first global composite view of Earth, shown in Figure 9.8. To the trained eyes of NASA scientists, the image revealed a multitude of weather events:

FIRST COMPLETE VIEW OF THE WORLD'S WEATHER

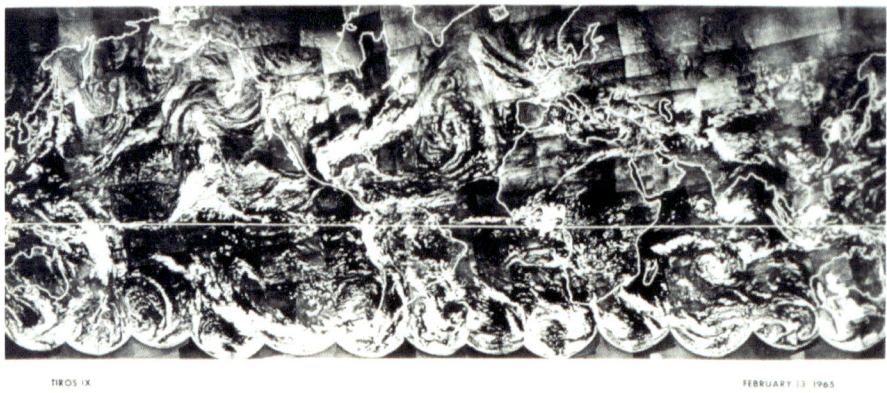

TIROS IX   FEBRUARY 13 1965

Figure 9.8 Mosaic of 450 images taken by TIROS-IX on February 13, 1965 to produce the first global composite view of Earth.
NASA

Figure 9.9 Natural color composite of images taken on February 13, 2015, 50 years after Figure 9.8, by the Visible Infrared Imaging Radiometer Suite (VIIRS) on the Suomi NPP satellite, named for satellite pioneer Verner Suomi.
NASA

A tropical storm over Ceylon and the southern tip of India, and another over the south Indian Ocean. In the lower right, a storm is approaching the southern coast of Australia. The thin band of clouds extending from central North Africa across the Red Sea to Saudi Arabia indicates the location of the jet stream. . .The remnants of an old storm are indicated by the comma-shaped cloud array over the North Atlantic Ocean.... A strong weather front is depicted by the clouds extending across the southeastern United States; another storm is moving into the northwestern United States from Canada.[4]

Fifty years later, as illustrated in Figure 9.9, satellites no longer produced blurry mosaics. Instead, computers now assemble full-color images of Earth daily, with a precision that was unimaginable in 1965. Notice the clarity with which the Red Sea, just to center right, is depicted and how weather patterns across the entire Earth can be viewed at once.

PART IV
SATELLITE SURVEILLANCE OF SEA AND ICE

10

Satellites and the Sea

Radar Detection

The true potential of satellites became apparent when scientists equipped them with specialized instruments to monitor Earth's seas, ice, and terrain. One of the first instrumental techniques used was radar, an acronym for Radio Detection and Ranging. Developed during World War II, it was essential for detecting the presence, direction, distance, size, and speed of planes, ships, and weather formations. Radar uses electromagnetic radiation at varying wavelengths, depending on what is being measured, with most applications falling within the frequency range of 1 GHz (billion cycles per second) to 100 GHz. This range places radar in the microwave region, situated between radio waves and infrared radiation in the electromagnetic spectrum, as illustrated in Figure 10.1. Microwaves have the advantage of traveling through the atmosphere with minimal obstruction. In a radar system, pulses of microwaves are emitted from a transmitter, reflect off an object of interest, and echo back to the radar system aboard the satellite. By measuring the time it takes for the echo to return, the distance, or "range," to the reflector can be calculated. The Doppler shift of the returning waves provides information about the speed and direction of an object's motion.

The land, ice, forests, and water below continuously receive the radar signal from the satellite altimeter and reflect it back. The larger the radar antenna on the satellite, the smaller the object that can be distinguished. Early satellites could carry only a relatively small antenna before they became too heavy to launch, which limited their resolution of small objects.

One of the first missions to conduct radar altimetry was the Geodynamic Experimental Ocean Satellite, or GEOS, which operated from 1975 to 1979. GEOS-3 mapped Earth's gravitational field, providing essential information for accurate land surveying. During this period, it also monitored changes in the extent of sea-ice cover, a vital indicator of climate change.

The 1978 Seasat mission (see Figure 10.2) marked a breakthrough in satellite observation as the first to use Synthetic Aperture Radar (SAR). The term refers to the technique by which an antenna can achieve the resolution of

Sentinels in the Sky. James Lawrence Powell, Oxford University Press. © James Lawrence Powell (2026).
DOI: 10.1093/9780197842850.003.0010

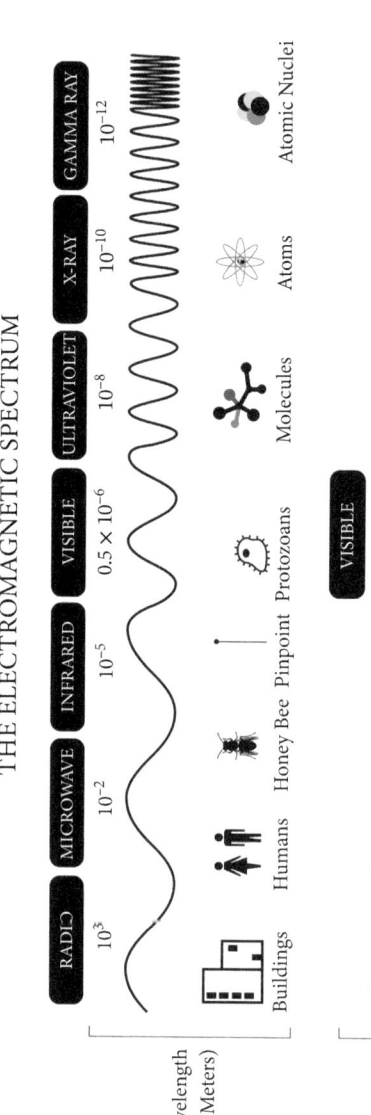

Figure 10.1 The Electromagnetic Spectrum. 1 megahertz (MHz) is equal to one million frequency cycles per second; 1 gigahertz (GHz) to one billion cycles per second. Solar radiation is mainly in the visible spectrum, while radiation emitted by Earth is in the infrared, which carbon dioxide absorbs preferentially.

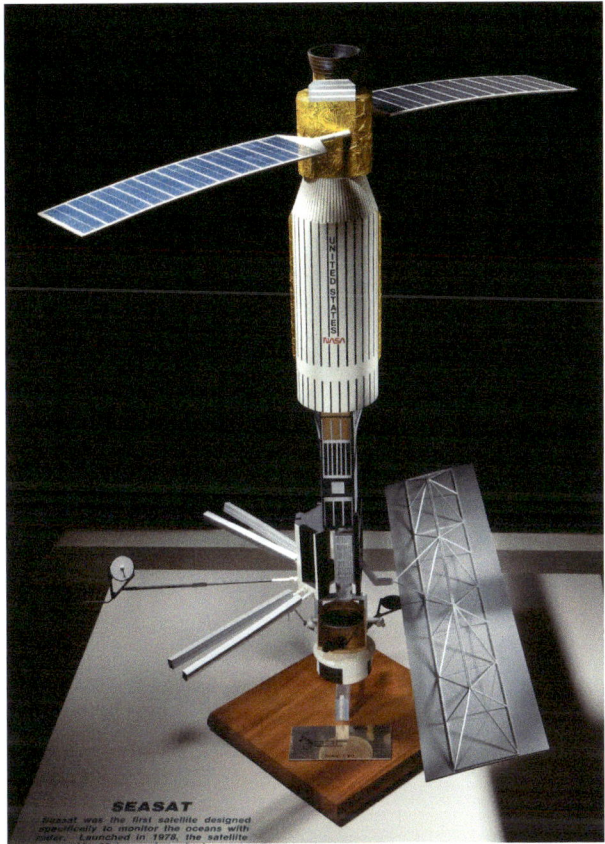

Figure 10.2 Seasat with twin solar panels at the top and the synthetic aperture radar antenna at the bottom right.
Wikimedia Commons

a much larger one. As the spacecraft travels along its orbit, it transmits a sequence of radar pulses and records the echoes reflected from the Earth's surface. By melding these successive returns, SAR effectively synthesizes a large antenna aperture, greatly improving spatial resolution. Crucially, SAR can operate day or night and through cloud cover, haze, or rain—capabilities that have made it an indispensable tool for Earth observation, disaster response, and environmental monitoring.

SAR is particularly effective for mapping and analyzing surface features, as it measures both the intensity and phase of the radar signal. The intensity indicates how strongly a surface reflects the radar, while the phase—which refers to the position of the wave within its cycle at a specific point in time—provides extremely precise distance measurements. This capability facilitates the creation of detailed topographic maps and 3D models. The applications of

SAR technology include monitoring glaciers, tracking deforestation, mapping urban development, detecting oil spills, and assessing flood-prone zones or areas affected by disasters.

A catastrophic short circuit, possibly caused by a design flaw in the electrical system or by an unexpected external event, such as a micrometeoroid impact or space radiation, caused Seasat to cease returning data 100 days after launch. But that brief period was enough to show the promise of satellite observation of the oceans. SAR's higher resolution enabled detailed mapping of ocean features such as wave patterns, currents, and wind speeds, as well as the area, thickness, and movement of sea ice. The data enhanced the understanding of ocean–atmosphere interactions and was crucial for weather forecasting and climate modeling. Seasat also made precise measurements of wave heights across the globe, which were used for both conducting naval operations and comprehending ocean dynamics.

Tide Gauges

For centuries, sailors relied on wooden stakes to track the tides, hoping to avoid disaster in shallow waters. While such knowledge was passed down anecdotally, the first tidal gauge was installed by the Dutch in 1675.

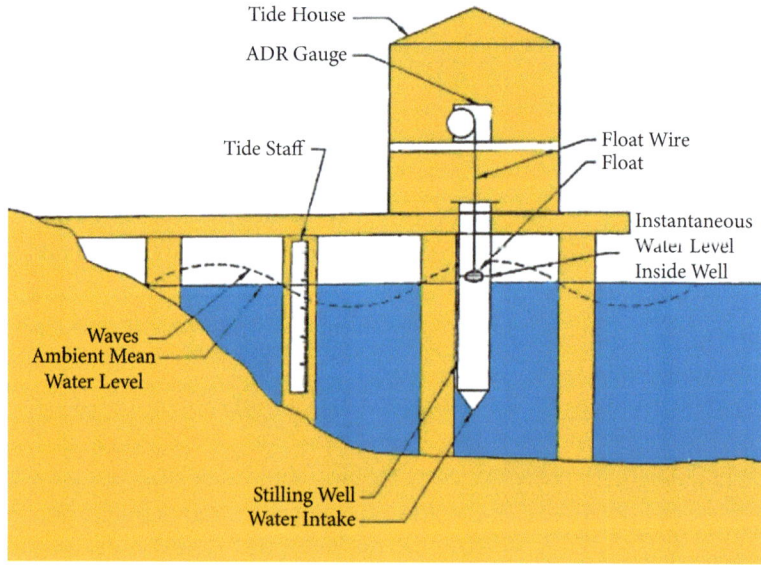

Figure 10.3 Schematic of a tidal gauge showing how it dampens the effect of waves.
NOAA

As illustrated in Figure 10.3, these early devices contained a float housed within a chamber or pipe connected to a larger body of water. The float would rise and fall with the changing water level. These chambers, known as stilling wells, dampened waves and surges, allowing accurate measurements that could otherwise be disrupted by turbulent water. By the late nineteenth and early twentieth centuries, tidal gauges had become widely used, providing critical data for navigation, coastal development, and planning. Tidal gauges remain in use today, not only to monitor sea-level changes but also to control gates or pumps in reservoirs.

Sea-Level Altimetry

As concerns over global sea-level rise intensified starting in the 1970s, so too did the need for a broader, more comprehensive view. Enter satellite altimetry—a technology that now enables scientists to monitor sea-level changes across the entire ocean, not just along the shorelines. To determine absolute sea level, scientists need to know the satellite's exact position in space, which is established using high-precision tracking systems such as GPS and satellite laser ranging. By combining the known orbital position with the measured distance to the ocean surface, sea-surface height can be calculated relative to a reference ellipsoid, providing a consistent global framework for monitoring sea-level change.

Satellites provide several key advantages over tide gauges in monitoring sea level. First, instead of being limited to specific locations where gauges happen to be installed, they offer uniform measurement across nearly the entire ocean surface. Second, even early satellite measurements achieved accuracy of a few millimeters. Third, satellites automatically gather and transmit data, enabling consistent long-term monitoring over decades. Figure 10.4 illustrates the change in global sea level since 1880, initially recorded by tide gauges and later measured by satellites.

In 1992, NASA and the French Centre National d'Études Spatiales launched the TOPEX/Poseidon satellite (see Figure 10.5) from the Guiana Space Center on the northeastern coast of South America The spacecraft carried two altimeters: one named TOPEX, for the Ocean Surface Topography Experiment, and the other named Poseidon, for the Greek god of the sea. Orbiting Earth every 112 minutes, TOPEX/Poseidon tracked rising sea levels with centimeter accuracy. The mission was designed to last three years, but the altimeters continued to function for an additional ten, while successor missions have maintained the record, with the results shown in Figure 10.6.

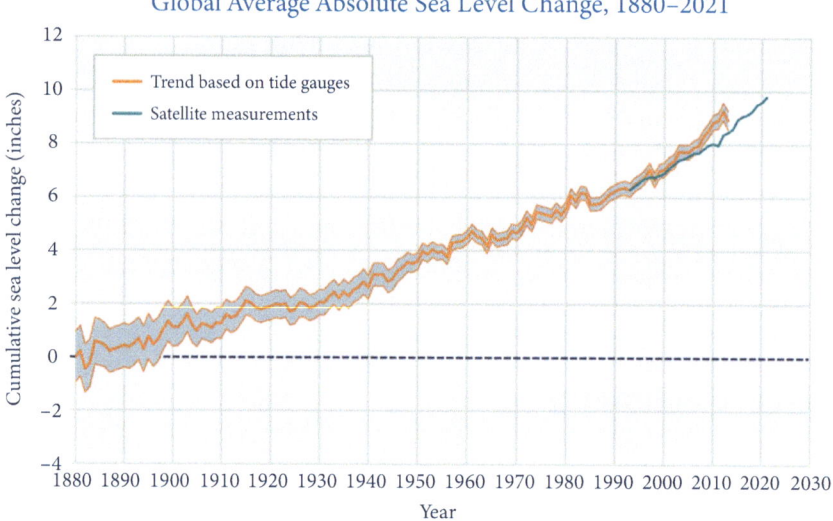

Figure 10.4 Sea-level rise, 1880–2021. The gray area shows the upper and lower bounds of the error in measurements from tidal gauges. The dark line represents the data from satellite altimetry, which began in 1992 and which has replaced tidal gauges. It is much more accurate, and, on this scale, it has no visible error bars.
Environmental Protection Agency

A close examination of Figure 10.4 and Figure 10.6 reveals that the sea-level curves show a subtle upward concavity. This is not an illusion but rather reflects a real phenomenon: sea-level rise is accelerating. In other words, with each passing year, the average increase in sea level is slightly greater than it was the year before.

The Intergovernmental Panel on Climate Change (IPCC) projects that global mean sea level will continue to rise throughout the twenty-first century and beyond, with the extent of the rise depending largely on future greenhouse gas emissions. Under midrange emissions scenarios, the IPCC estimates sea-level rise of approximately 0.5 to 1 m by 2100. A 1-m increase would have profound consequences: coastal land and infrastructure would be lost, storm surges would reach farther inland, and saltwater intrusion would threaten freshwater supplies.

Millions of people in vulnerable areas—especially small island nations, low-lying deltas, and major coastal cities—could face displacement. Home-owners in coastal areas are already finding it increasingly difficult to obtain financing and insurance, leaving rebuilding efforts affordable only to the wealthy. Coastal ecosystems such as wetlands and mangroves would suffer

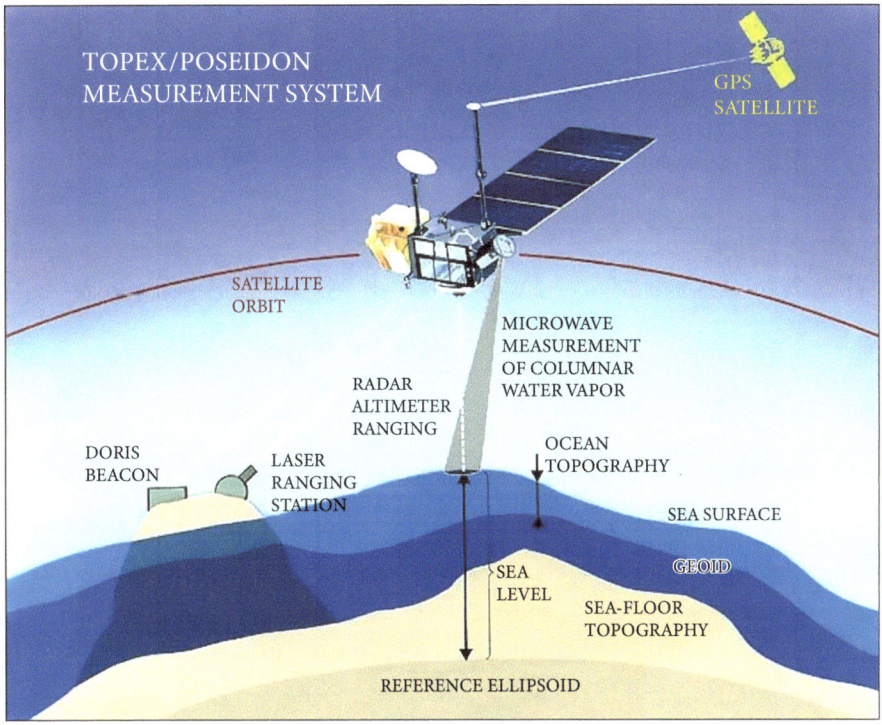

Figure 10.5 Diagram of the TOPEX/Poseidon measurement system, showing how satellite radar altimetry, ground-based laser tracking, and GPS positioning are integrated to measure ocean topography and sea level with remarkable precision, critical for understanding ocean circulation, climate change, and sea-level rise. eoPortal

major disruptions, with cascading effects on biodiversity and the natural barriers they provide.

This is not a distant or speculative scenario—it is already unfolding in many parts of the world.

Scatterometry

Instruments called scatterometers emit microwave pulses toward the ocean, land, or other surfaces and analyze the angle and intensity of the reflected signals. Over the sea, wind speed affects surface roughness, which in turn influences how much radar energy is scattered. Calm water reflects the signal like a mirror, with little scattering, whereas rougher seas—driven by

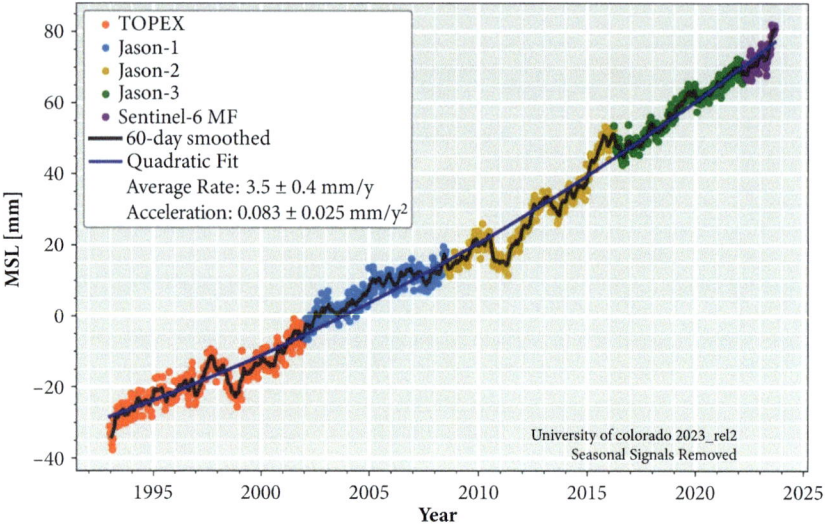

Figure 10.6 Satellite measurement of sea level since 1992.

University of Colorado

stronger winds—produce more extensive scattering. By analyzing these patterns, scatterometers provide real-time data on wind speed and direction, wave conditions, hurricanes, and shifting sea ice. Their ability to detect wind direction from the asymmetric scatter pattern makes them indispensable for ocean wave modeling, weather forecasting, and selecting optimal locations for nearshore wind farms.

Beyond wind measurements, scatterometry measured from the land surface has many diverse applications. Variations in soil moisture influence radar backscatter, making scatterometers valuable for hydrology and drought monitoring. Different types of land cover, such as forests, grasslands, and urban areas, have unique scattering characteristics, allowing scatterometers to differentiate between terrains. At high latitudes, the difference in backscatter between sea ice and open water allows scatterometers to monitor polar ice cover, even in cloudy conditions or darkness. Continuous observation delivers essential data on sea-ice dynamics, a crucial indicator of climate change.

Scatterometry is also employed to monitor and respond to various natural and humanmade hazards. During a storm surge, higher wind speeds generate larger waves and rougher sea surfaces, leading to distinctive backscatter patterns. Scatterometer measurements of wind direction and speed indicate where the storm surge is heading and provide early warnings. Scatterometry offers the advantages of covering vast areas, functioning in all weather conditions, and continuously monitoring as satellites regularly revisit the same locations. Below is a more detailed description of some of scatterometry's most significant applications.

Oil spills are all too familiar, with notable instances such as the Deepwater Horizon explosion and oil spill in the Gulf of Mexico in 2010, the largest in U.S. history. In the past, oil spills have been tracked from ships or planes, but these methods have obvious disadvantages compared to the broader field of view and continuous coverage offered by satellites. When oil spreads over water, it reduces surface tension, smoothing out waves and creating calmer conditions. This, in turn, lessens the backscatter signal. The movement of oil spills is primarily driven by the surface currents, which can be tracked by satellite. Hindcasting suggests that this mapping technique, had it been available at the time, would have accurately predicted the trajectory of the Deepwater Horizon oil spill.

Icebergs pose a persistent threat to maritime navigation—and not only in polar waters. When the *Titanic* struck its fatal iceberg in 1912, the ship was near the latitude of Massachusetts, a stark reminder of how far icebergs can drift. Unlike sea ice or open water, icebergs exhibit a markedly higher backscatter intensity in radar imagery, making themdetectable by satellites equipped with SAR, such as the European Space Agency's (ESA's) Sentinel-1. However, identifying and tracking icebergs in near-real time requires analyzing vast amounts of remote sensing data, a task that is both storage- and computation-intensive.

This is where the Google Earth Engine (GEE) has become invaluable. Introduced in 2010 to detect illegal deforestation, GEE has evolved into a powerful cloud-based platform capable of processing petabytes of satellite and geospatial data. It integrates imagery from missions such as Landsat, Sentinel, and MODIS (Moderate Resolution Imaging Spectroradiometer), along with climate and environmental datasets. By providing built-in tools for data access, processing, and visualization, GEE allows researchers to monitor icebergs efficiently, even under cloudy skies or during polar night. Its capacity to distinguish sea ice from open water using radar data has made it a vital tool in modern iceberg surveillance and maritime safety.

In another new development, researchers have successfully used artificial intelligence to map the extent of Antarctic icebergs, cutting down the analysis time from several minutes to well under one second.[1] The French oceanographic data center CERSAT offers archival maps showing the locations of small icebergs (less than 3 km in length).[2]

Satellite Bathymetry

The first attempt to measure ocean depth—or bathymetry—was made by the crew of the British *Challenger* expedition from 1872 to 1876.[3] They dropped a weighted and knotted line over the side until it hit the bottom and then counted the number of knots that had played out. Early in the twentieth century, echo sounders began to be used to emit sound waves and measure the time it took for the reflected signal to return. Since the speed of sound in seawater is approximately 1500 m/s, the depth of the seafloor can be calculated from the travel time of an acoustic signal. For example, if the echo returns after three seconds, the depth would be about 2250 m, calculated as (1500 m/s × 3 s) ÷ 2. Echo sounders run continuously, displaying the result on a device that draws a vertical profile of the seafloor. In remote areas where few ships travel, the coverage from echo sounders is sparse, but satellite observations are filling this gap.

By the 1990s, satellite radar altimetry had become sensitive enough to detect slight variations in sea-surface height that reflect underlying seafloor topography. Figure 10.7 illustrates how the method works. Imagine a mountain on the ocean floor—a "seamount." Its mass exerts a localized gravitational pull that slightly alters Earth's gravity field, drawing water toward it and creating a small outward bulge on the sea surface. Satellite altimeters detect this bulge, and from it, scientists can infer the presence and approximate size of the seamount below. Currently, a seamount needs to be roughly a mile (about 1.6 km) high to produce a sea-surface anomaly large enough to be reliably detected.

When combined with data from echo sounding and other techniques, satellite bathymetry has significantly contributed to mapping the seafloor,

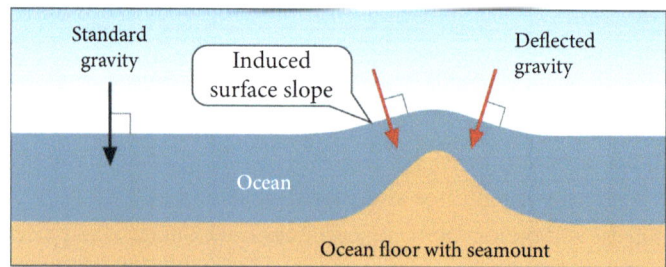

Figure 10.7 The gravitational pull of a seamount adds to Earth's overall gravity, producing small slopes in sea level. Synthetic Aperture Radar can detect these slopes and reveal seafloor topography.
NOAA/NESDIS/STAR

especially large-scale features, as shown in Figure 10.8. While detailed high-resolution maps rely primarily on ship-based measurements, the combined data are available on platforms such as Google Earth and GEBCO.net (the General Bathymetric Chart of the Oceans).

Satellite mapping of seafloor topography has turned up thousands of previously uncharted seamounts and related features. They include:

The discovery of a new tectonic plate, the Mammerickx Microplate in the Indian Ocean east of Madagascar, named in honor of Jacqueline Mammerickx, a pioneer in seafloor mapping. Research on the microplate has helped establish the timing of the collision between the Indian and Eurasian tectonic plates, which formed the Himalayas.

The use of high-resolution seafloor maps to study subduction zones, where one tectonic plate descends beneath another, generating earthquakes and volcanism. For example, after the 2011 Tohoku earthquake in Japan, bathymetric surveys revealed significant seafloor changes that contributed to the devastating tsunami.

The identification of the source of a massive pumice raft that erupted in July 2012 near the Kermadec Islands, about 1,000 km northeast of New Zealand's North Island. Before it dispersed, the pumice raft was estimated to be 3.5 m thick. Its origin remained uncertain until satellite

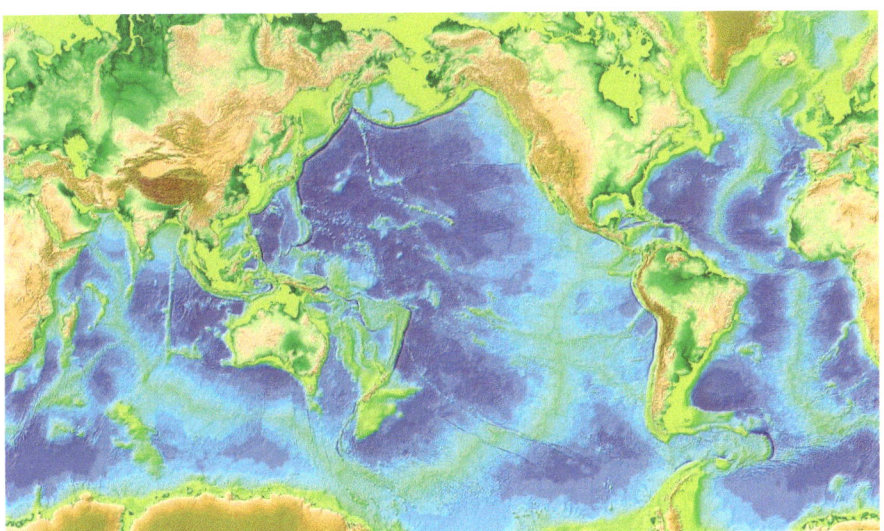

Figure 10.8 Global Seafloor Topography derived from satellite altimetry and shipboard depth soundings.

NOAA

bathymetry revealed it had erupted from Havre Seamount, previously thought to be inactive.

The mapping of fault structures off the coast of Southern California, including the San Diego Trough Fault Zone and the Palos Verdes Fault. High-resolution bathymetry has provided vital data on their slip rates, lengths, and potential seismic hazards. As discussed in Chapter 7, these two faults are part of a complex system of "Borderland faults" that may accommodate 15 to 20 percent of the motion between the Pacific and North American plates.

LIDAR

The acronym LIDAR stands for Light (or Laser) Detection and Ranging. Like radar, it measures the time it takes for a signal to travel to an object and reflect back, but by using light waves instead of radio waves, LIDAR achieves much higher resolution. This precision makes it especially valuable for generating detailed 3D maps. The robotic helicopter *Ingenuity* used LIDAR to fly over and map the surface of Mars. Remarkably, LIDAR has even been used to map the ocean floor. A specially designed Airborne Bathymetric LIDAR emits light pulses toward the ocean's surface, which, if the water is sufficiently clear, can penetrate to the seafloor and return through the water to the sensor. Two pulses are returned: one from the water's surface and one from the seafloor, something like the measurement of snow depth discussed in Chapter 7. LIDAR uses the time difference between the arrival of these pulses to calculate the water depth and, therefore, seafloor topography.

Advances in LIDAR technology have extended its operational range beyond 200 km, making it suitable for use from satellites in low earth orbit. A satellite LIDAR system can produce a global map of Earth's surface with a resolution of 30 m. Although the method is currently too expensive for routine use, technological advances are likely to allow expanded use of LIDAR.

Rivers and Lakes

The principle behind satellite altimetry—measuring the height of a water surface relative to a fixed reference—applies not only to oceans but also to freshwater lakes, rivers, wetlands, and floodplains. While rising sea level is a key long-term indicator of climate change, inland water bodies respond more rapidly and visibly to climatic shifts, acting as sentinels of change on land.[4]

Unlike the ocean, whose level rises gradually at about 3.5 mm/year, lakes and rivers can shrink or surge over seasons or even weeks, with immediate consequences for ecosystems, agriculture, and human consumption. For example, Utah's Great Salt Lake has contracted to its smallest recorded volume; Lake Mead has fallen to levels last seen in the 1930s when it was filling; and upstream, Lake Powell now holds just over one-third of its capacity. The Great Salt Lake, a remnant of Ice Age Lake Bonneville, is natural. Lakes Mead and Powell are humanmade reservoirs formed by the Hoover and Glen Canyon dams, which generate hydroelectric power and provide critical water supplies to the American Southwest. Monitoring these inland water levels by satellite offers early warning of hydrological stress and highlights yet another facet of the global climate crisis.

Worldwide, people depend on reservoirs for water and power much more than on underground aquifers (which are rapidly depleting) or on desalinating seawater (which is costly, environmentally damaging, and will not come close to fulfilling the world's freshwater needs). By the beginning of this century, around 58,000 dams taller than 15 m had been constructed, covering a total reservoir surface area of nearly 300,000 sq km. They provide flood control, hydropower generation, and water for irrigation and drinking.[5] But as shown by the three examples above, these humanmade reservoirs are more vulnerable than people may have realized.

Beginning with TOPEX/Poseidon, a series of missions have provided satellite altimetry alongside photography, enabling scientists to measure the levels of inland seas, rivers, lakes, wetlands, and reservoirs. The Hydroweb database, created at the LEGOS laboratory (Laboratoire d'Études en Géophysique et Océanographie Spatiales) in Toulouse, France, now encompasses data for approximately 150 large lakes worldwide and about 20 major rivers, as illustrated in Figure 10.9.[6] Its database spans over ten years of records.

The Congo Basin, the world's second-largest river basin, plays a vital role in the global climate system, similar to that of the Amazon Basin. In 2014, researchers using data from the European Space Agency's ENVIronmental SATellite (ENVISAT), launched in 2002 as part of the Earth Observation Program, demonstrated that satellite altimetry could accurately measure water levels in the Congo Basin's streams, even in the narrow upstream sections. This capability is essential for enhancing flood prediction and monitoring long-term climate change.

The remote Tibetan Plateau is home to over 1,000 lakes that exceed 1 km^2 in size. They are located at an average altitude of 4,000 m, making access difficult. Known as the "Third Pole," this lofty plateau serves as the headwaters of several of the world's largest river systems, including the

Figure 10.9 Hydroweb Measurement Stations as of November 2020.
Hydroweb

Brahmaputra, Ganges, Indus, Mekong, Salween, Yangtze, and Yellow Rivers. These waterways support more than 1.3 billion people—nearly one-fifth of the global population—by providing water for drinking, irrigation, and power generation.

Over the past fifty years, the Tibetan Plateau has experienced a warming rate of approximately 0.36°C per decade, nearly double the global average. This recent rise in temperature has sped up the melting of the plateau's 37,000 glaciers, leading to an increase in the level of its lakes by an average of about 20 cm each year. As warming continues, the glaciers will further shrink, and once most of the ice has melted, the volume of the plateau's rivers, lacking a source to replenish the lost meltwater, will fall.

11

GRACE

Geologists have long recognized that when the great continental ice sheets reached their maximum extent around 18,000 years ago, global sea levels stood approximately 120 m lower than today, as vast volumes of ocean water were locked away in glacial ice. Subsequent changes in Earth's orientation in space and distance from the Sun led to warming that melted the ice sheets and raised sea level, but a vast amount of ice still remains in the caps covering the Arctic, Greenland, and Antarctica. The melting of this ice and the resulting rise in sea level has become a critical issue for humanity's future, making it essential to monitor the status of these giant ice masses.

The Gravity Recovery and Climate Experiment

Reflecting this concern, a 1997 report by the National Research Council (NRC) singled out the Antarctic and Greenland ice sheets as the greatest potential contributors to future sea-level rise. The report stressed that tracking changes in ice mass through gravity measurements, in conjunction with laser altimetry, would significantly improve our understanding of ice sheet dynamics. The NRC based its recommendation on the principle that variations in Earth's mass distribution—such as the loss of ice from melting glaciers, shifting groundwater, or ocean currents—produce measurable changes in the planet's gravitational field. This insight laid the groundwork for the eventual development of the Gravity Recovery and Climate Experiment (GRACE). Among the many satellite-based innovations in Earth science, GRACE stands out for its ingenuity, groundbreaking discoveries, and wide-ranging interdisciplinary impact.

To detect changes in Earth's gravitational field, the GRACE mission uses two identical satellites flying in tandem at an altitude of approximately 500 km in a near-polar orbit, separated by about 220 km. The satellites continuously measure the distance between them using a highly precise microwave ranging system, capable of detecting changes on the order of a few microns—a micron being one-millionth of a meter, or about one-hundredth the thickness of a human hair.

Sentinels in the Sky. James Lawrence Powell, Oxford University Press. © James Lawrence Powell (2026).
DOI: 10.1093/9780197842850.003.0011

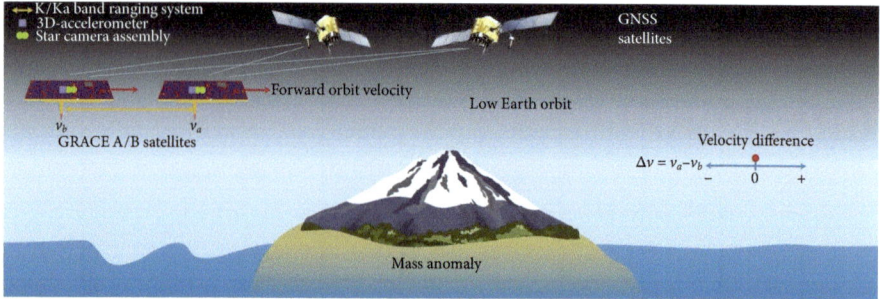

Figure 11.1 Schematic of the GRACE twin satellites approaching a high-gravity mass, from left to right. Far from the mass, both satellites maintain the same velocity, but as the lead satellite nears, it accelerates, resulting in an increase in the distance between the two satellites, which can be measured with great precision.

Byron D. Tapley et al., "Contributions of GRACE to Understanding Climate Change," *Nature Climate Change* 5, no. 5 (April 2019): 358–69, https://doi.org/10.1038/s41558-019-0456-2.

As illustrated in Figure 11.1, when the lead satellite approaches an area with stronger gravitational pull—such as a mountain range or a large ice sheet—it experiences an increased pull and accelerates slightly, increasing the gap between itself and the trailing satellite. Once past the area, the lead satellite slows down while the trailing satellite, now approaching the region, undergoes the same velocity increase. This continuous variation in satellite separation directly reflects the distribution of mass below.

By analyzing these data, scientists generate detailed maps of Earth's gravitational field, applying algorithms to model gravitational pull and detect anomalies. GRACE surveys the entire planet every thirty days, producing monthly updates to the global gravity model, making it an essential tool for monitoring water storage, ice loss, and mass redistribution on a planetary scale.

Of course, there are complications, not least because almost all of Earth's gravitational pull originates from its interior rather than from features on the surface. That said, the deep crust and mantle remain stable for long periods, while the pull of water at the surface—oceans, rivers, glaciers, ice caps, and groundwater—varies on a much shorter timescale. This allows the short-term effects to be distinguished from the long-term pull of gravity.

The two GRACE satellites stem from an international collaboration led by NASA's Jet Propulsion Laboratory and the German Aerospace Center in Potsdam. The twin satellites were launched on March 17, 2002, from the Plesetsk Cosmodrome in northern Russia, the site of Korolev's early R-7 rocket launches, aboard a Soviet SS-19 rocket. GRACE was initially intended to operate for five years but functioned for fifteen years before being decommissioned in October 2017.

Vanishing Ice

GRACE provided the first direct measurement of changes in the mass of the great polar ice caps. The results were shocking. Within two years, GRACE measurements revealed that both Greenland and Antarctica were losing ice at an alarming rate. Since then, the loss has continued unabated. From 2002 to 2017, as illustrated in Figure 11.2, Greenland lost 258 Gt of ice per year, while Antarctica lost 137 Gt per year. GRACE data indicate that approximately 60 percent of the mass loss from Greenland results from increased melting due to rising surface temperatures, while the remainder comes from an increase in the amount of ice calving off Greenland as icebergs have entered the sea and melted there.

The Arctic is warming at four times the global average primarily due to feedbacks driven by the loss of sea ice. As ice melts, it exposes darker ocean water, which absorbs more solar energy than reflective ice, accelerating warming, furthering ice loss, and so on. This feedback loop, combined with changes in atmospheric and oceanic circulation, amplifies regional warming well beyond the global average. Since the 1980s, Arctic sea-ice volume has decreased by at least 70 percent. The Arctic Ocean is expected to face its first ice-free day within the next few decades, according to climate models.

The amount of snow Antarctica receives is linked to low atmospheric pressure over the Amundsen Sea, shown just below the Antarctic Peninsula in the right frame of Figure 11.2. This low-pressure system, in turn, is associated with the El Niño cycle, a climate pattern characterized by periodic warming of sea-surface temperatures in the central and eastern tropical Pacific Ocean, which disrupts normal atmospheric circulation. This illustrates how the climate in one area can affect and be affected by others, even those as isolated and distant as Antarctica.

GRACE is so sensitive that it can detect ice loss not only from massive ice sheets but also from smaller mountain glaciers, even in remote regions where direct measurements are scarce. These regions include glaciers in Alaska, the Canadian Arctic, Patagonia, and Svalbard, a Norwegian archipelago above the Arctic Circle. One striking example is China's vast Tien Shan range, which is home to thousands of glaciers.[1] Their melting plays a vital role in sustaining surface and groundwater supplies for farmers in Northwest

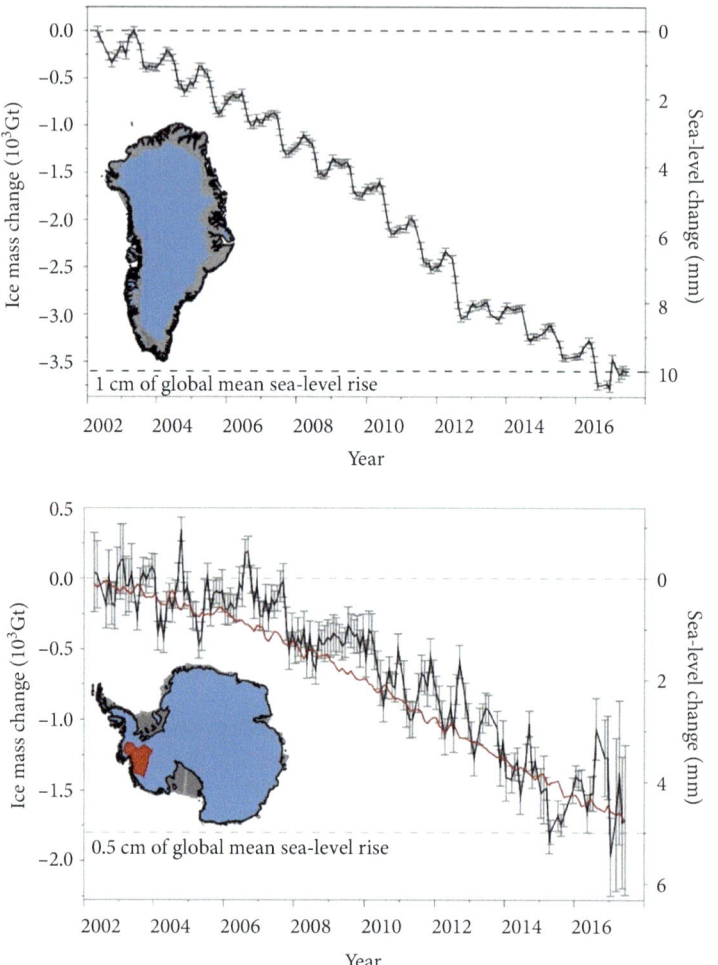

Figure 11.2 GRACE observations of mass change of the polar ice caps between April 2002 and June 2017.

Tapley et al., 2019.

China, particularly those cultivating cotton. The region receives only about 20 cm of precipitation annually, requiring glacial meltwater to support over three-quarters of China's cotton production. However, satellite data from GRACE and other missions reveal that glacier coverage in the Tien Shan shrank by approximately 18 percent between 1961 and 2012, mainly due to intensified summer melting. While melting has temporarily boosted water availability to the farmers, continued ice loss without replenishment will ultimately shrink the region's supply of meltwater and curtail or end cotton production.

Groundwater

The volume of ice caps and freshwater bodies in lakes and rivers at Earth's surface can be measured or estimated, but much of the near-surface water is groundwater, out of sight. Groundwater is renewable in that it is replenished by rainfall; however, in many countries, including the United States, it is being depleted at a much faster rate than can be replaced. Consequently, in numerous areas, the water table is dropping. As water is added to or removed from underground aquifers, the resulting change in mass slightly alters the local gravitational pull. GRACE's twin satellites track these variations. By removing the influence of surface water and atmospheric moisture using other data sources, scientists can isolate and map changes in groundwater storage over time. A study using GRACE conducted by researchers at the University of California, Irvine, found that from 2002 to 2017, of the world's thirty-seven largest groundwater aquifers, approximately one-third have been seriously depleted due to both climate change and overuse.[2] The most extreme loss occurred in arid Arabia, where over 60 million people depend on aquifers for their water.

Soil Moisture and Drought

The primary tool for drought monitoring and prediction in the United States is the U.S. Drought Monitor.[3] This service provides weekly maps that summarize drought conditions based on information from various sources, including ground measurements and local reports from climatologists. Before 2011, the Drought Monitor did not include data on water storage and groundwater. However, NASA then began incorporating satellite data on groundwater from GRACE and other missions. This has resulted in more accurate and near real-time drought maps, as illustrated in Figure 11.3. Note how well the results from GRACE match up with the traditional drought monitor map.

Flood Forecasting

Unlike drought, which operates on a timetable of months and seasons, floods arrive in a matter of hours and days. Therefore, flood prediction requires the kind of real-time data that satellite remote sensing can best provide. The degree of wetness of a watershed predetermines how it will respond to precipitation, allowing GRACE-derived wetness indicators to be used in

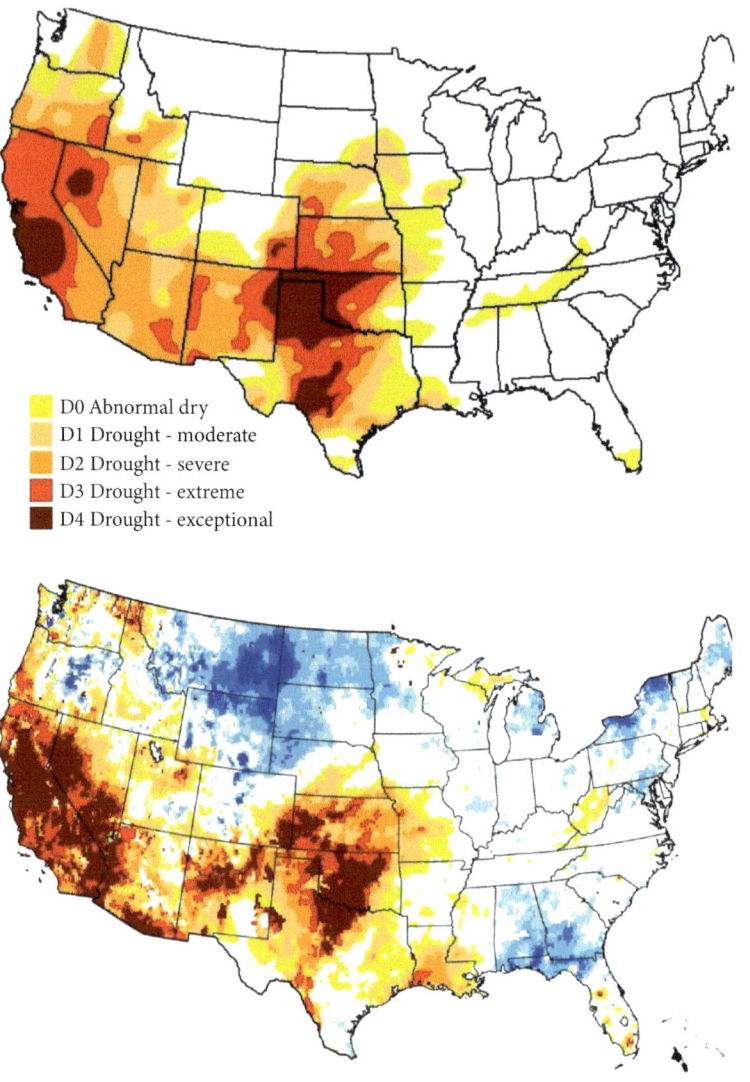

Figure 11.3   The top frame displays the U.S. Drought Monitor for May 20, 2014. Compare it with the bottom frame from the previous day, which illustrates soil moisture from GRACE, available automatically within 24 hours. The dark areas indicate the probability of a location being drier than at the same time of year during the record period from 1948 to the present.

Modified from Tapley et al. (2019)

flood forecasting. One study found that data from GRACE allowed the predisposition of a watershed to flooding to be predicted five to eleven months in advance. Case studies of floods in the Columbia and Indus River basins confirmed this result.[4]

Wildfires

Before the advent of GRACE, no long-term, spatially comprehensive record of soil moisture existed to support reliable wildfire forecasting. To explore the potential of GRACE data for this purpose, researchers at NASA and the U.S. Forest Service analyzed historical fire records alongside GRACE-derived soil moisture patterns.

The analysis revealed a key distinction: smaller wildfires were more likely to follow wetter-than-average pre-fire seasons, when increased vegetation growth provided abundant fuel. In contrast, the largest and most destructive fires tended to erupt in areas where soils were already dry, which promoted the rapid spread and intensification of fire. If this relationship holds under continued observation, GRACE data could become an essential tool in forecasting wildfire risk, especially as climate change increases both the frequency and severity of these events.

Sea-Level Change and Ocean Bottom Pressure

To monitor ocean temperature and salinity, data from GRACE are combined with measurements from the Argo network, a global system of floating buoys. Argo tracks thermal expansion of sea water by measuring changes in ocean density, whereas GRACE detects mass redistribution. By subtracting the thermal expansion contribution from the total sea-level rise recorded by satellite altimeters, scientists can isolate the portion caused by melting ice, thereby improving estimates of climate-driven sea-level change.

Another surprising application of GRACE data is the measurement of changes in ocean bottom pressure. But how can satellites orbiting hundreds of kilometers above Earth detect subtle variations at the seafloor? The key lies in gravity. As the height of the water column increases, so does the mass of water above the seabed—raising the pressure at the bottom and slightly increasing the local gravitational pull. GRACE's ultrasensitive instruments can detect these minute changes in gravity, allowing scientists to calculate variations in ocean bottom pressure from measurements in space.

Deep ocean currents are largely driven by density variations in the water column, which can be inferred from ocean bottom pressure. These currents, in turn, shape global ocean circulation. Density differences arise from changes in temperature and salinity, altering the vertical pressure gradient that drives deep water movement. Since these currents regulate heat transport, climate patterns, and even sea-level rise, GRACE data provide a powerful tool for tracking large-scale shifts in ocean circulation, improving models of Earth's climate system.

Solid Earth Changes

GRACE is also used to measure postglacial rebound—technically known as glacial isostatic adjustment—a process in which Earth's crust slowly rises in response to the melting of the massive ice sheets. As the weight of the ice diminishes, the underlying crust, once depressed, begins to rebound. The explanation lies in the behavior of the upper mantle, which, over geological timescales, can flow like a viscous fluid. This allows the crust to "float" and adjust its elevation according to differences in density. The result is a dynamic balance that helps maintain a relatively uniform gravitational field across Earth's surface, with lighter continental crust and mountain ranges standing high, while denser oceanic regions sit lower. In the 1880s, American geologist Clarence Dutton coined the term *isostasy*—from the Greek for "equal standing"—to describe this equilibrium of floating crustal blocks atop the deformable mantle.

Ice is heavy, as anyone who has tried to lift a block of it knows. At the height of the Ice Age, much of the Northern Hemisphere was covered by a sheet of ice up to 3 km thick. The immense weight of these sheets pushed the viscous mantle below out of the way, depressing the land surface. When the ice melted, the land rebounded and mantle material flowed in at depth to replace that moving upward, something like a mattress that slowly regains its initial form after you get up. In the SE Hudson Bay area, for example, as measured by GRACE, this ongoing isostatic adjustment can amount to 11 mm/yr.

GRACE Follow-On

The wealth of essential data provided by GRACE made a compelling case for its continuation after decommissioning in 2017, leading to the launch of the "GRACE Follow-On" mission the following year. To measure the distance between the pair of satellites, GRACE-FO employed a new instrument called the Laser Ranging Interferometer, which enhanced the precision of intersatellite measurements by a factor of 20. Unfortunately, the critical microwave instrument failed, delaying the start of data collection until the end of January 2019. GRACE-FO has a design lifetime of five years, but we can hope it may, like its predecessor, operate for much longer.

PART V
ENERGY AND TEMPERATURE

12

The Earth's Energy Imbalance

Energy Received = Energy Emitted

Over the long run, which in the case of our solar system means billions of years, Earth and any solid body in space must match the energy received from the Sun with the energy emitted back into space. A planet that emits less energy than it receives from the Sun—that retains more, in other words—would gradually heat up, causing its surface to melt. Conversely, a planet that emits more energy than it receives would cool, leading its surface to freeze.

When Earth absorbs more energy from the Sun than it radiates back into space, the excess is stored in the atmosphere, oceans, and land. As a result, surface temperatures rise, climate patterns shift, polar ice melts, sea levels rise, oceans warm and acidify, and ecosystems are disrupted. Conversely, when Earth emits more energy than it receives, global temperatures fall, winters grow harsher, ice sheets and glaciers advance, sea levels fall, and growing seasons shorten. What sets modern global warming apart is its unprecedented pace—unfolding not over geological epochs, but within a single human lifetime. This reality underscores the urgent need to monitor Earth's energy balance continuously—a task for which satellites are uniquely well suited.

Earth's energy balance begins with the amount of solar radiation it receives, defined by a quantity known as the solar constant—the flux of solar energy, measured in watts per square meter (W/m^2), that reaches the upper atmosphere on a surface oriented perpendicular to the Sun's rays. (A watt equals one joule of energy per second, roughly the power used by a small light-emitting diode nightlight.) This influx of solar energy drives nearly all processes at Earth's surface, from atmospheric circulation to photosynthesis. In contrast, heat escaping from Earth's interior contributes only a minute fraction—several thousand times less than solar input. For two reasons, the energy actually received per square meter at Earth's surface is, on average, only about one-quarter of that entering the top of the atmosphere. Two main factors account for the reduction. First, at any given moment, only half of Earth is illuminated by the Sun, effectively halving the average energy received. Second, the angle at which sunlight strikes Earth varies with

Sentinels in the Sky. James Lawrence Powell, Oxford University Press. © James Lawrence Powell (2026).
DOI: 10.1093/9780197842850.003.0012

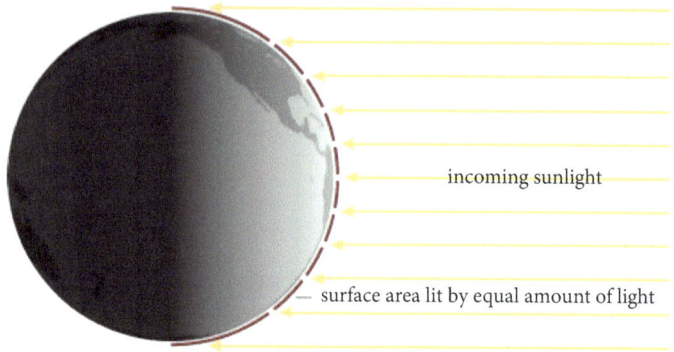

Figure 12.1 From the equator to the poles, the Sun's rays meet Earth at smaller and smaller angles, and the sunlight gets spread over larger and larger surface areas.
NASA

latitude: rays arrive nearly perpendicular at the equator, but they strike at increasingly oblique angles toward the poles, spreading the same amount of energy over a larger surface area, as illustrated in Figure 12.1. These combined geometric effects reduce the average solar energy reaching the surface to approximately one-fourth of the solar constant,

Scientists have long recognized the significance of the solar constant. One of the first to measure it was French scientist Claude Pouillet, who in 1838 obtained a value of 1,228 W/m^2. The accuracy of his measurements, as well as those of others prior to the satellite era, was limited by weather and atmospheric conditions. Beginning with Nimbus 7 in 1978, a series of satellite missions have attempted to pin down the value of the solar constant. The most precise measurement to date was made in 2019, during a period of minimal sunspot activity, yielding a value of 1,361.6 ± 0.3 W/m^2—only about 10 percent higher than Pouillet's nineteenth-century estimate. To visualize this, the solar constant is roughly equivalent to the energy of fourteen 100 W light bulbs shining on every square meter at the top of Earth's atmosphere. The energy reaching the surface is one-fourth of that amount, approximately 340 W/m^2. A portion of that energy is reflected, so the net solar energy received at the surface is around 240 W/m^2.

Figure 12.2 illustrates Earth's energy budget, showing the balance between incoming and outgoing radiation as observed by satellites, as of 2009. The most crucial measure for assessing climate change is the imbalance between these two, represented by a single value known as Earth's Energy Imbalance (EEI) and shown on the figure as "net absorbed." This metric converts the complex interplay of atmospheric processes and feedback mechanisms into a single indicator of planetary heating.

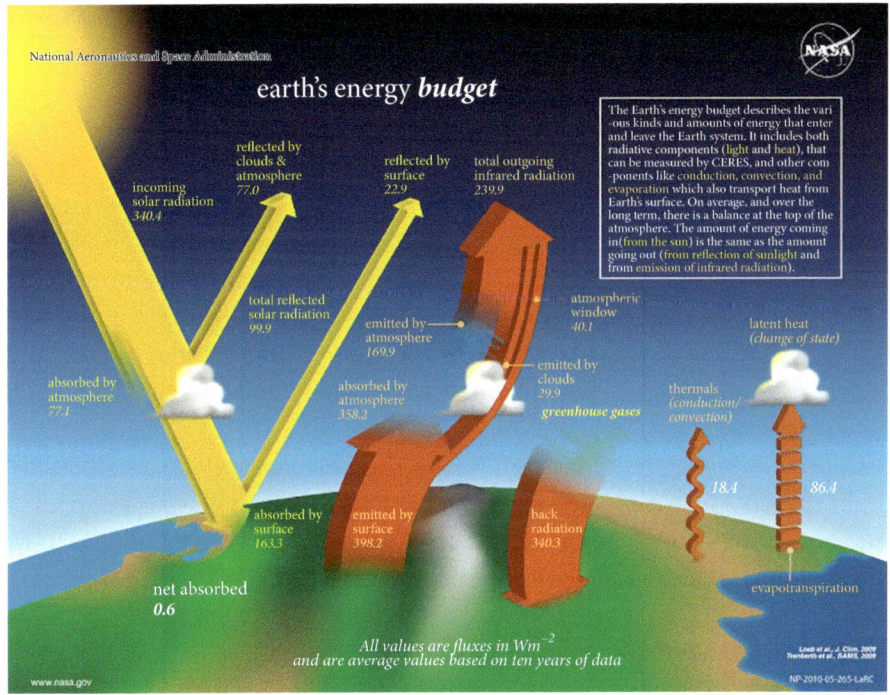

Figure 12.2 Earth's energy budget as measured by satellite.

NASA

Incoming solar radiation was measured at 340.4 W/m², while outgoing radiation—comprising reflected shortwave (77.0 W/m²), atmospheric window (22.9 W/m²), and longwave radiation (239.9 W/m²)—totaled 339.8 W/m². (The "atmospheric window" refers to the portion of longwave (infrared) radiation emitted by Earth's surface that escapes directly to space without being absorbed by greenhouse gases.)

Accordingly, the EEI in 2009 was +0.60 W/m²—clear evidence that Earth is gaining energy and is thus warming. By 2019, the EEI had increased to 0.90 ± 0.15 W/m². While this figure may seem small, its impact is amplified by Earth's enormous surface area: 5.1×10^{14} m². To grasp the significance, consider that 0.90 W/m² sustained over the entire planet adds up to an energy gain of roughly 1.5×10^{22} joules in a single year—most of it stored in the oceans. By comparison, total global energy consumption by humanity in 2019 was about 6×10^{20} joules. In other words, Earth retained twenty-five times more heat from the energy imbalance than all the energy used by human civilization in that year. Such a persistent imbalance cannot continue without triggering profound and potentially dangerous consequences. As the great geochemist Wallace Broecker once remarked, "The climate system is an angry beast and we are poking it with sticks."[1]

ERBS

The Earth Radiation Budget Satellite (ERBS) was a groundbreaking NASA mission aimed at studying Earth's radiation budget and its contribution to climate change. Launched on October 5, 1984 from the Space Shuttle Challenger, ERBS collected data on the absorption, reflection, and emission of solar energy across the atmosphere, helping scientists refine climate models and assess the effects of greenhouse gases. One of ERBS's key contributions was providing long-term observations of Earth's albedo—the fraction of sunlight reflected by the planet's surface and clouds. The satellite also measured outgoing longwave radiation, which helps quantify the greenhouse effect. By tracking these variables, ERBS confirmed that rising greenhouse gas concentrations were shifting Earth's energy balance, leading to a net gain in heat—an early indication of manmade global warming.

ERBS was designed for a two-year lifetime, but it remained operational for nearly four decades, a testament to its robust design. Throughout this period, it faced the familiar challenges of aging but it persevered. In 1990, the scanner failed; in October 1993, a computer memory error occurred; and in 1998, one of the two telemetry units ceased functioning. A year later, an equipment failure halted solar measurements. By the mission's end, only one of the five stabilizing gyros was still operational, and the fuel tank bladder had failed. After nearly two decades in orbit, ERBS reentered Earth's atmosphere on January 8, 2023, burning up over the Bering Sea. Any senior citizen can sympathize.

CERES

ERBS paved the way for successor satellite missions such as CERES (Clouds and the Earth's Radiant Energy System), which was designed to determine the net effect of clouds on the planet's climate—a long-standing challenge for climate models, since clouds play a dual role in Earth's energy system: They reflect incoming solar radiation, cooling the planet, but they also trap outgoing infrared radiation, contributing to warming. Their overall impact depends on factors such as the altitude, thickness, and composition of clouds, as well as on complex feedback mechanisms that either amplify or dampen climate change. By providing high-resolution data on cloud properties—including height, thickness, and optical characteristics—CERES enhanced understanding of these interactions, reducing uncertainty in climate projections.

Aerosols

Aerosols are suspensions of microscopic solid particles or liquid droplets in air or another gas. Whether natural or humanmade, aerosols have effects similar to clouds in that they both reflect incoming solar radiation and absorb heat rays rising from Earth's surface, complicating the assessment of the net impact of aerosols on global climate.

Natural aerosols include fine volcanic ash, water droplets from fog and mist, and dust lifted into the atmosphere from drought-stricken and arid regions. Human-made aerosols arise from a wide range of sources, including industrial emissions, vehicle exhaust, pesticide application, power generation, agricultural activity, burning of waste, and smoke from both natural and human-caused forest fires. Additionally, the condensation of gases such as sulfur dioxide from burning dirty coal creates aerosols. One of the most common sources is fly ash—fine particles expelled from coal-fired power plants during combustion—from industrial furnaces, illustrated in Figure 12.3.

In the spring of 1991, ERBS recorded the eruption of Mount Pinatubo in the Philippines. By volume of material ejected, it was the second-largest volcanic explosion of the twentieth century, surpassed only by the 1912 blast from Novarupta in Alaska's Katmai Peninsula. Mount Pinatubo expelled over 10 km^3 of volcanic material into the atmosphere, approximately ten times the amount released during the 1980 eruption of Mount St. Helens in Washington State. All three of these volcanoes—Pinatubo, Novarupta, and Mount St. Helens—are located along the Ring of Fire that encircles the Pacific Ocean, where one tectonic plate descends (subducts) under another. The descending oceanic plate transports water and other volatile materials to depths where the pressure is sufficient to trigger an explosion that lifts hot magma to the surface.

The eruption of Mount Pinatubo injected an estimated 20 million tons of sulfur dioxide (SO_2) into the stratosphere, where it reacted with water vapor to form a veil of sulfate aerosols. These minute particles spread around the globe, reflecting incoming solar radiation back into space and temporarily cooling Earth's surface. From its vantage point in orbit, the Stratospheric Aerosol and Gas Experiment II (SAGE II) aboard Nimbus-7 monitored the evolution of the volcanic plume, documenting in detail how the aerosols

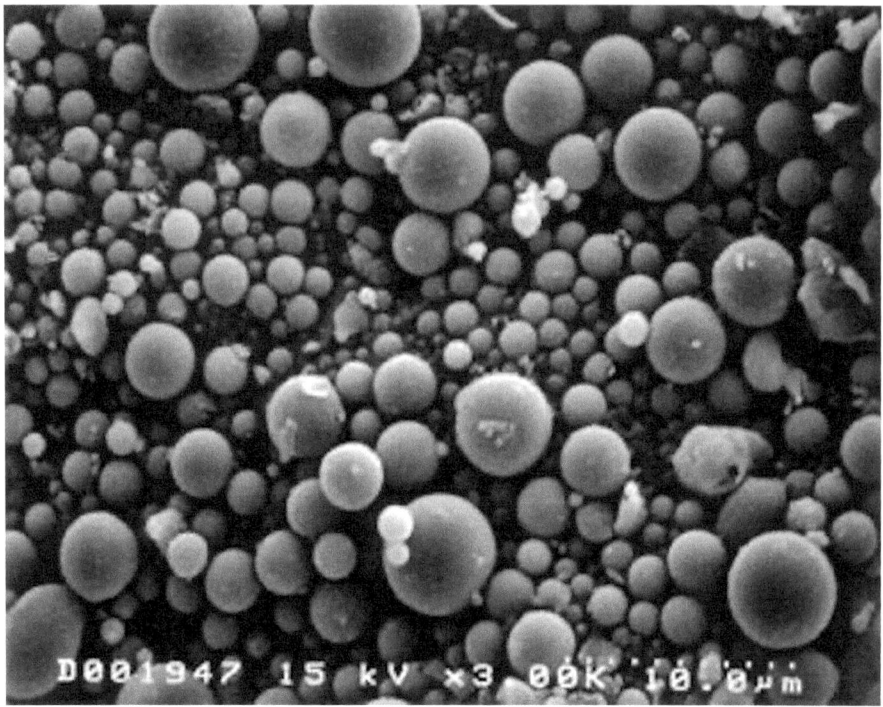

Figure 12.3 Fly ash as photographed by a scanning electron microscope. Magnified about 2,000 times.
Wikimedia Commons

dispersed through the upper atmosphere and contributed to short-term global cooling.

The event provided a rare, large-scale natural experiment, helping scientists refine climate models by enhancing their understanding of how aerosols affect radiation balance, atmospheric chemistry, and ozone depletion. Additionally, the Pinatubo eruption contributed to a short-term acceleration of ozone loss, as sulfate particles facilitated chemical reactions that destroyed stratospheric ozone, particularly in polar regions.

The idea that volcanic aerosols reflect incoming sunlight and cool our planet has a long history, with Benjamin Franklin being the first to suggest it. Franklin served as the United States Minister to France from 1778 to 1785. During the summer of 1783, while he was in Paris, a persistent "dry fog" enveloped much of Europe, including the "City of Lights." This phenomenon dimmed sunlight and was associated with a noticeable drop in temperature. Franklin observed and documented these unusual atmospheric conditions, noting their potential impact on climate. He was aware that the Icelandic volcano Laki had erupted in 1783 in what was one of the largest and most deadly eruptions in recorded history. Laki emitted vast clouds of sulfur dioxide and other aerosols, leading to a famine that led to the deaths of many people and their animals. It is a tribute to the resilient Icelanders that they recovered from this unprecedented disaster and went on to build the modern country that we know today.

In 1784, at a meeting of the American Philosophical Society in Philadelphia, Franklin proposed that the Laki eruption in Iceland had released vast quantities of airborne particles and gases, producing the mysterious "dry fog" observed across Europe and North America. Two other major eruptions in the nineteenth century had similarly far-reaching climatic effects. In 1815, Mount Tambora, on the Indonesian island of Sumbawa, erupted catastrophically, triggering the "year without a summer." This was followed in 1883 by the eruption of Krakatau in the Sunda Strait, between Java and Sumatra, which also caused significant global cooling. The effects of Krakatau peaked in 1884, making it one of the coldest years recorded since global temperature measurements began in 1880.[2]

NASA scientist James Hansen and his colleagues used the 1991 Mount Pinatubo eruption to predict how sulfur dioxide aerosols would affect global temperatures, providing a real-world test for their computerized climate models. They forecast that temperatures would decline slightly but would recover within a few years. In January 1997, Hansen's team evaluated their prediction's accuracy and found that by 1992–1993, global temperatures had dropped 0.6°C, but by 1995 had returned nearly to pre-eruption levels. This confirmation led them to analyze four other major volcanic eruptions of the twentieth century. Each showed a similar pattern of short-term cooling followed by recovery.

This research has fueled discussions about geoengineering, specifically the idea of injecting sulfur dioxide or other aerosols into the atmosphere to increase Earth's albedo, reflecting more sunlight and temporarily cooling the planet, thus offsetting global warming. This strategy might be feasible in the short term, but its long-term viability remains in doubt, as it would require global coordination, sustained funding, and political will over centuries—all while avoiding unintended consequences. The biological and environmental risks of injecting large quantities of sulfur into the atmosphere are unknown but could include disruptions to global precipitation patterns, damage to the ozone layer, acid deposition, and detrimental effects on ecosystems. Why bet on geoengineering when we have a virtually risk-free alternative: a transition to renewable energy sources, including nuclear power?

The Ozone Hole

One crucial, yet largely invisible, component of the atmosphere is ozone—a colorless, toxic gas composed of three oxygen atoms (O_3). Formed when ordinary oxygen (O_2) is exposed to electrical discharges or ultraviolet (UV) radiation, ozone plays an essential protective role. High in the stratosphere, it acts as a natural shield by absorbing the bulk of the Sun's harmful UV rays. Without this barrier, life on the surface would face elevated risks of skin cancer, cataracts, and other health problems, while marine ecosystems and agricultural productivity would also suffer serious harm.

Antarctica provides an ideal natural laboratory for monitoring ozone levels, owing to its exceptionally clean air, which offers an unobstructed view of atmospheric chemistry. The region's extreme cold fosters the formation of polar stratospheric clouds, which play a key catalytic role in ozone depletion. At the same time, the Antarctic polar vortex acts as an atmospheric barrier, isolating the air mass over the continent. This containment allows scientists to study ozone loss in a relatively controlled setting, while also gaining insights into global atmospheric trends.

In 1985, three scientists from the British Antarctic Survey reported that from the 1970s to the early 1980s ozone levels above Antarctica had declined by over 30 percent. They used a ground-based instrument that measured incoming UV radiation to estimate ozone concentration.[3] The SAGE II satellite subsequently confirmed and tracked this decline, which became known as the Ozone Hole.

The cause of the ozone hole remained a mystery until researchers pinpointed chlorofluorocarbons (CFCs), such as Freon, as the primary culprits. Once commonly used in aerosol sprays, refrigerants, and foam-blowing devices, CFCs can persist in the atmosphere for up to 100 years. When exposed to UV radiation, CFCs release chlorine atoms, which catalyze the breakdown of ozone molecules, resulting in an overall decline in atmospheric ozone. The large chemical companies that sold CFCs initially resisted this conclusion but ultimately had to concede to the scientific evidence.

To address public concern over the dangers of CFCs and similar substances, on September 16, 1987, twenty-four countries along with the European Community adopted the Montreal Protocol on Substances that Deplete the Ozone Layer. This historic treaty aimed to phase out the production and consumption of ozone-depleting substances, including CFCs and chemicals known as halons. President Ronald Reagan, not typically recognized as an environmentalist, stated:

> The Montreal protocol is a model of cooperation. It is a product of the recognition and international consensus that ozone depletion is a global problem, both

in terms of its causes and its effects. The protocol is the result of an extraordinary process of scientific study, negotiations among representatives of the business and environmental communities, and international diplomacy. It is a monumental achievement.[4]

The Montreal Protocol has functioned as promised: atmospheric concentrations of ozone-depleting substances have declined, the ozone layer is gradually recovering, and the protocol has also helped mitigate climate change by targeting compounds that are potent greenhouse gases. Perhaps its most important legacy is the demonstration that the international community can respond to a global environmental threat with unity, scientific integrity, and enforceable action. The Paris Agreement on Climate—today's aspirational counterpart—aims to confront the even greater challenge of global warming, but its long-term success remains uncertain. The contrast between the two serves as both a warning and a call to action: collective agreements can work, but only if they are backed by commitment and accountability by all nations.

Planetary Heat

Another approach to tracking Earth's energy budget measures planetary heat uptake—a phenomenon related to, yet conceptually different from, Earth's Energy Imbalance (EEI). While EEI represents the difference between absorbed solar radiation and outgoing thermal radiation emitted to space, planetary heat uptake captures where this excess energy actually goes within the Earth system. This accumulated energy manifests as measurable changes: warming oceans, melting ice sheets and glaciers, rising sea levels, and increasing atmospheric and land temperatures. The oceans absorb more than 90 percent of this surplus energy, while the remaining fraction drives atmospheric warming, ice loss, and surface heating.

Planetary heat uptake arguably provides a more fundamental measure of global warming than surface temperature alone. Surface temperatures fluctuate with weather patterns, shifting ocean currents, and other temporary factors, whereas planetary heat uptake reflects the steady accumulation of energy throughout the entire Earth system—making it a robust indicator of long-term climate change.

A useful comparison is to consider planetary heat uptake as a vital sign for Earth's climate system—much like monitoring a patient's metabolism. In the same way that physicians evaluate respiratory function through oxygen consumption and carbon dioxide production, tracking planetary heat uptake reveals the fundamental health of our climate system. Surface temperature

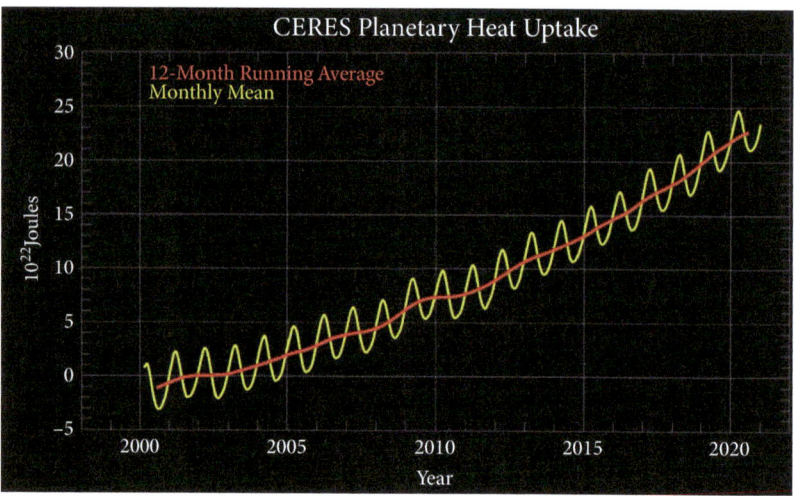

Figure 12.4 As shown by data from the CERES satellite mission, Earth has stored excess heat every year since 2005. The oscillations stem from the seasonal cycle, which is positive between October and April and negative between May and September.

NASA

readings can vary considerably from year to year, influenced by natural cycles and regional variations. Planetary heat uptake, however, cuts through this noise to reveal the steady, underlying accumulation of energy that drives long-term climate change—providing a more reliable indicator of the planet's energetic imbalance.

As shown in Figure 12.4, despite nations' promises to reduce greenhouse gas emissions, the pulse of our planet is not only increasing; it is accelerating dangerously, with no end in sight. Any thoughtful person examining this chart should ask themselves what could reverse this trend and what might happen if it is not reversed.

Conclusion

This chapter tells a story of disruption—of a planet taking in more energy than it releases, banking the surplus in oceans, atmosphere, and ice sheets. The consequences have moved beyond theoretical predictions into daily reality. A constellation of satellites—ERBS, CERES, GRACE, SAGE, and others—now monitors Earth's energy budget with extraordinary precision, observing the delicate balance that makes our world habitable. These instruments do more than document climate change mechanics; they measure its velocity and direction. The data leave no room for ambiguity: Earth is accumulating heat at an accelerating pace, validating decades of climate science and highlighting the narrow window for action. Earth's energy balance represents more than an abstract measurement it is the fundamental rhythm of our planetary system, and satellites have given us the tools to monitor it. How we respond to their findings will shape the trajectory of civilization for generations.

13

Land- and Sea-Surface Temperatures

The concept of using satellites to measure Earth's surface temperature is nearly as old as Earth-orbiting satellites themselves. The TIROS II meteorological satellite, launched from Cape Canaveral on November 23, 1960—just three years after Sputnik—carried radiometers that measured infrared (heat) radiation emitted by Earth's surface. The data clearly indicated a cold front extending across northern Florida and reaching altitudes of nearly 30,000 feet. Since then, as discussed in Chapter 9, weather satellites have measured land-surface temperature as a crucial element of weather forecasting.

Landsat

In 1965, William T. Pecora, a visionary director of the U.S. Geological Survey, noted the information contained in the images taken by cameras aboard the Gemini and Mercury crewed missions and proposed using a remote sensing satellite to monitor features of our planet's surface. After typical Washington bickering, the Landsat program launched its first satellite on July 23, 1972. Landsat 1 was equipped with an infrared and visible-light scanner designed by Virginia Norwood, an MIT graduate and engineer at Hughes Aircraft Company. It was followed by Landsat 8, which launched on February 11, 2013 and within one year had more than 1.3 million image downloads. Landsat 8 produced one of the first complete views of the contiguous United States, as shown in Figure 13.1. Initially, Landsat images had to be purchased, but in 2008, the U.S. Geological Survey made them freely available. This led to the widespread use of Landsat data in agriculture, forestry, land-use planning, and environmental monitoring, spawning tens of thousands of research articles.

The latest addition to the Landsat program, Landsat 9, launched on September 27, 2021, extending a nearly five-decade legacy of continuous Earth observation. The satellite carries two primary instruments: the Operational Land Imager 2 (OLI-2), which captures visible, near-infrared, and

Sentinels in the Sky. James Lawrence Powell, Oxford University Press. © James Lawrence Powell (2026).
DOI: 10.1093/9780197842850.003.0013

Figure 13.1 This Landsat 8 image, taken on its first anniversary, shows one of the first satellite views of the contiguous United States. The satellite captures images in swaths, which are assembled into the composite shown.

NASA

shortwave infrared light across nine spectral bands, and the Thermal Infrared Sensor 2 (TIRS-2), which measures land-surface temperature in two thermal bands. From its orbital altitude of 705 km, Landsat 9 captures imagery at 30-m resolution for most spectral bands, with 15-m resolution for panchromatic imagery and 100-m resolution for thermal measurements.

Operating in tandem with Landsat 8, Landsat 9 ensures complete global coverage every eight days—each satellite imaging the entire Earth every sixteen days. This latest satellite represents an evolutionary advancement over its predecessor, featuring enhanced radiometric resolution that enables detection of subtler variations in surface conditions. This improvement proves particularly valuable for monitoring darker landscapes such as dense forests and coastal waters, where previous instruments struggled to distinguish fine details. Together, these twin satellites maintain the continuity of the world's longest-running terrestrial observation program, providing essential data for tracking environmental change, agricultural productivity, urban expansion, and countless other applications that depend on consistent, long-term Earth observation.

Ground-based weather stations, first established in the 1850s, employ thermometers positioned a few meters above ground level—roughly at face height—to directly measure air temperature. Satellites, by contrast, measure the temperature of Earth's surface itself—the temperature you would feel if you touched the ground. Instead of direct temperature measurement, satellites detect the intensity of thermal radiation emitted from the surface and convert these readings to temperature values. This indirect approach introduces several challenges: atmospheric dust, water vapor, and aerosols all interfere with the radiation signal and must be corrected for accurate readings.

Satellite measurements face additional complications from the space environment itself. Orbital drift occurs when gravitational forces from the Moon, Sun, and planets gradually alter a satellite's altitude, inclination, and timing—changes that can affect temperature measurements if not properly accounted for. The instruments themselves degrade over time as exposure to cosmic radiation and the harsh space environment affects sensor sensitivity and accuracy. These inherent limitations make regular recalibration and sophisticated correction algorithms essential for maintaining the precision of satellite temperature records over the years or decades that these instruments operate.

Ground-based stations offer highly accurate, long-term temperature records, but their global distribution is uneven. Coverage is especially sparse in remote regions such as the polar areas and large deserts—just where Earth's most extreme temperatures often occur. In contrast, satellites provide continuous global temperature observations and can sample across multiple atmospheric layers, from the surface through the troposphere and into the stratosphere. This vertical profiling capability is vital for both weather forecasting and long-term climate analysis.

Satellites employ both microwave and infrared sensors to measure temperature across different wavelengths, with each technology offering distinct capabilities. Microwave sensors provide all-weather monitoring by penetrating cloud cover to reach the surface, maintaining consistent measurements regardless of atmospheric conditions. These instruments can also probe beneath the surface, proving essential for applications like determining sea-ice thickness and measuring soil moisture content in agricultural regions. Their ability to see through clouds makes them particularly valuable for maintaining continuous temperature records in persistently cloudy regions like the tropics and polar areas.

Infrared sensors complement microwave measurements by delivering superior spatial resolution and more detailed surface temperature maps, though they cannot penetrate clouds and they are affected by the diurnal cycle. Despite these limitations, infrared technology remains the backbone of operational temperature monitoring due to its lower cost and widespread deployment across weather satellite fleets. The combination of both sensor types creates a comprehensive observing system—microwave instruments ensuring continuity through all conditions, while infrared sensors provide the fine-scale detail needed for applications ranging from urban heat island stud-ies to agricultural monitoring. Together, these complementary technologies enable scientists to construct robust global temperature datasets that capture both broad patterns and local variations in Earth's thermal state.

The Measure of Global Warming

With the threat of global warming looming, tracking Earth's temperature is one of the most critical tasks in science. Figure 13.2 shows the change in global average temperature as calculated by six different research organizations, pri-marily based on weather station data. Each center employs a slightly different method, but all yield nearly identical results. Figure 13.3 presents tempera-ture data mainly from satellites, showing how anomalously hot 2023 and 2024 were.

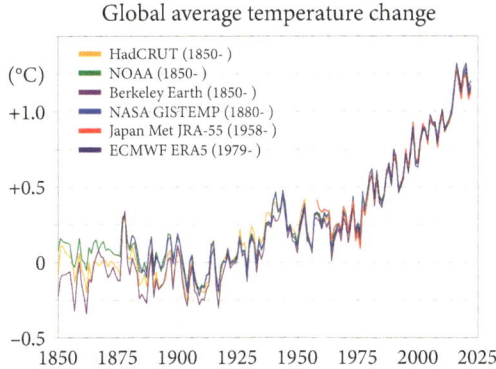

Figure 13.2 Global average temperature change as measured by six independent scientific organizations shows close agreement, underscoring the reliability of the data and showing clear evidence of global warming.

Wikimedia Commons

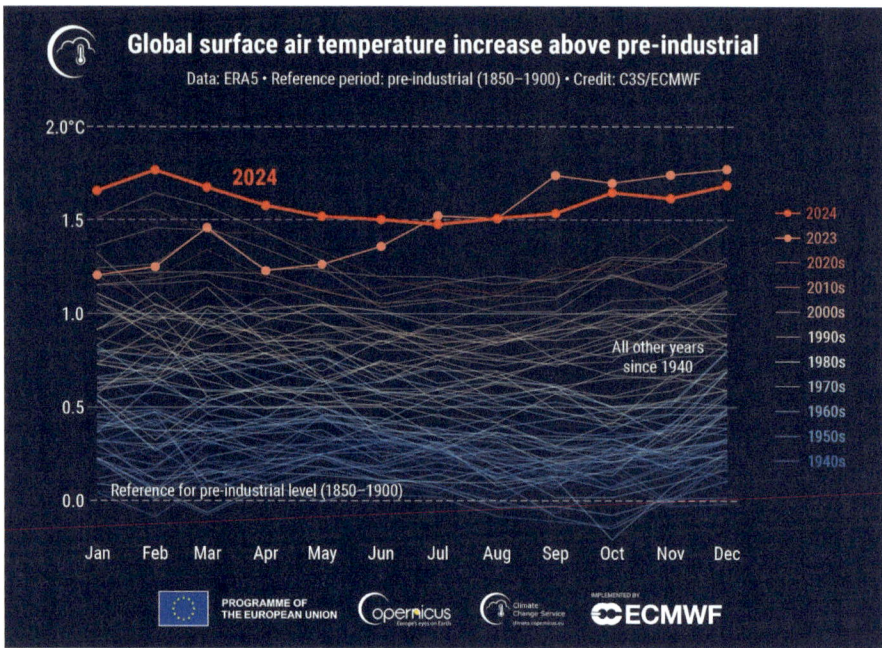

Figure 13.3 Global surface air temperature increase (°C) above the preindustrial average (1850–1900) for each month from January 1940 to December 2024. The data come from a variety of sources, including several dozen satellites, which confirm and enhance the record from ground measurements. According to the Copernicus program, July 22, 2024 marked the hottest day ever recorded.
Copernicus

Warming Troposphere, Cooling Stratosphere

Figure 13.4 illustrates the zones of the atmosphere from the surface upward. Understanding the temperature of each and its variations over time is crucial for assessing Earth's energy budget and patterns of greenhouse warming.

The temperatures of the troposphere and lower stratosphere have been recorded using radiosondes suspended from weather balloons since the 1930s, but satellites have transformed the coverage, frequency, and accuracy of these measurements. Infrared measurements began with the TIROS satellites and continued in the mid-1960s with NASA's second-generation Nimbus program. These instruments gauge the amount of radiation emitted by atmospheric gases, including carbon dioxide, but they cannot pass through cloud cover. Nimbus-5, launched in 1972, was equipped with a microwave-sounding unit that could penetrate clouds and gather data regardless of the weather. As mentioned, microwave spectrometers do not measure

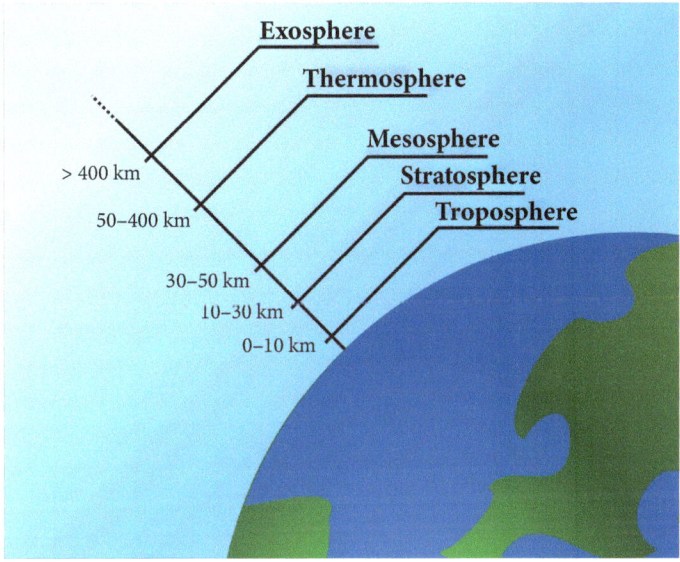

Figure 13.4 Zones of the atmosphere.
Wikimedia Commons

temperature directly; instead, they assess the intensity of radiation at specific wavelengths characteristic of different atmospheric gases. The warmer the gas, the more microwave energy it emits. However, microwave detectors have lower resolution than infrared ones. Combining the two maximizes the advantages of each. These early measurements, along with those that followed, have supplied decades of temperature data for the troposphere and stratosphere.

Those who deny that global warming is caused by human activity often argue that the Sun is to blame, not greenhouse gases. But this claim is false. Satellite and ground-based observations show that the Sun's energy output has remained essentially constant over the past several decades, during which time Earth's surface and lower atmosphere have warmed significantly. If the Sun were the driver of modern climate change, we would expect both the troposphere and the stratosphere to warm—especially the latter, which lies closer to space and receives more direct solar radiation.

Figure 13.5 shows the vertical temperature structure of Earth's atmosphere under natural, preindustrial conditions, unaffected by human influence. However, satellite measurements over recent decades reveal a different trend: the troposphere is warming while the stratosphere is cooling—a reversal of the natural vertical pattern and a distinctive prediction of greenhouse gas theory.

Troposphere

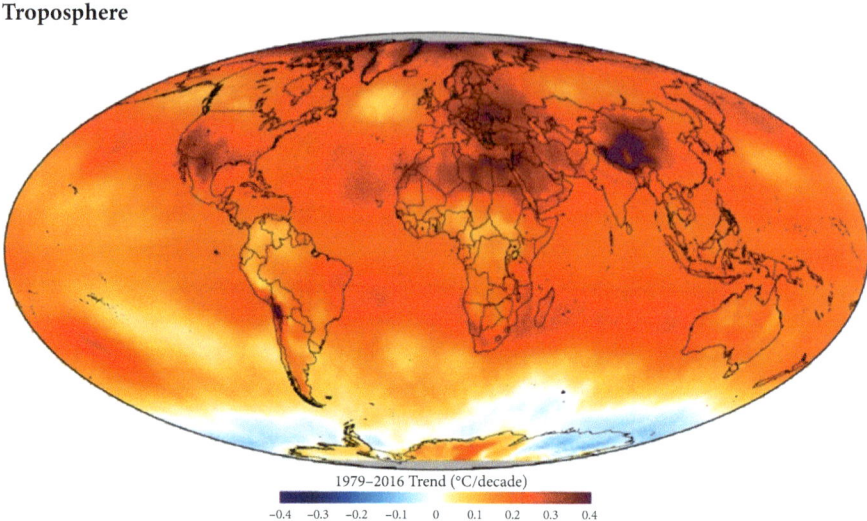

Lower Stratosphere

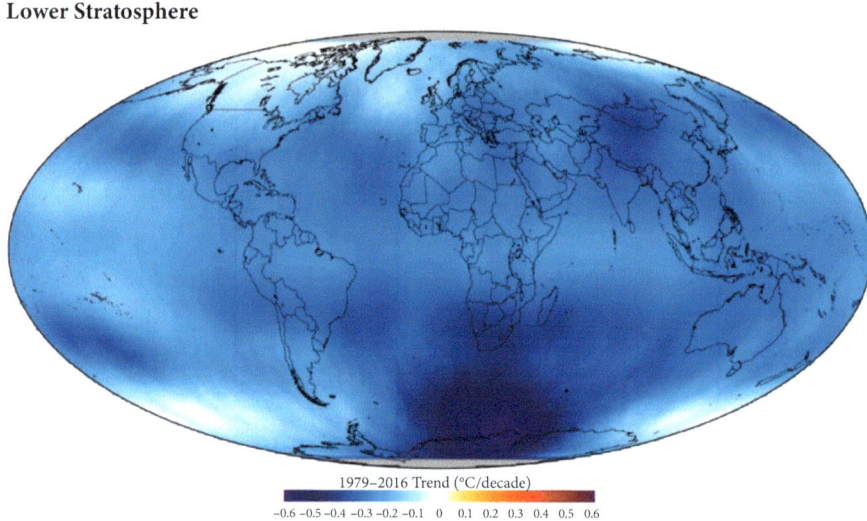

Figure 13.5 Average vertical temperature profile of Earth's atmosphere. The troposphere cools with altitude, while the stratosphere warms due to absorption of ultraviolet radiation by ozone. The mesosphere cools again, and the thermosphere heats dramatically at higher altitudes. Although the stratosphere normally shows a warming trend with height, satellite observations over recent decades reveal that it is cooling over time—a key signature of anthropogenic climate change. This cooling, combined with tropospheric warming, is a distinctive "fingerprint" of greenhouse gas forcing, not explainable by natural variability or increased solar activity alone.
Wikimedia Commons

This trend has occurred because greenhouse gases such as carbon dioxide and methane absorb outgoing infrared radiation from Earth's surface and re-radiate it within the troposphere, trapping heat and raising temperatures in the lower atmosphere. As more energy is retained below, less reaches the stratosphere, depriving it of the infrared radiation that once helped warm it. The result is stratospheric cooling, despite its natural warming trend with altitude due to ozone's absorption of ultraviolet light.

Urban Heat Islands

In today's warming world, a staggering one-third of humanity endures extreme heat for over twenty days each year, a figure sure to rise if global warming continues unchecked. Cities, with their dense infrastructure and active environments, exacerbate this heat, creating "urban heat islands" that can be up to 6°C warmer than surrounding rural areas. This phenomenon is driven by the heat-absorbing steel, asphalt, and concrete; pollution from vehicles and factories; and limited green space.

Traditional temperature measurement methods, such as weather stations, are often positioned in green urban parks or open airport spaces, typically away from city centers. As a result, they tend to underestimate the urban heat island effect, which is more pronounced in densely built-up areas. In contrast, satellites equipped with infrared and microwave spectrometers can provide comprehensive data to assess urban heating risks. These sensors not only capture temperature variations across entire metropolitan areas but also distinguish different land uses, such as industrial zones, residential neighborhoods, and parks, enabling more accurate analysis of urban heat distribution.

Sun-synchronous, polar-orbiting satellites like Landsat are especially suited for monitoring urban heat. For example, Landsat-8 crosses the equator at approximately the same time each day, ensuring consistent lighting conditions for its images. Over a sixteen-day cycle, it systematically covers the entire Earth, capturing swaths about 185 km wide, as seen in Figure 13.1.

Data from Landsat-8 and similar satellites have been essential for studying urban heat islands in major cities like Beijing, Hanoi, Milan, Paris, Philadelphia, and Stuttgart. The last-named stands out for its proactive strategy to combat urban heat and climate change. Stuttgart has adopted a comprehensive plan to protect green spaces and maintain fresh air corridors, receiving international recognition and the title of "the coolest city in the world"—a phrase that both serves as a literal reference to temperature and gives a nod to the city's innovative urban planning. The success of these initiatives is evident: a 2022 study using Landsat data revealed that while the overall land-surface temperature in the larger Stuttgart region rose by 2.5°C between 2004–2008 and 2016–2020, the temperature within the urban heat island itself decreased by 1.4°C.

The Paris summer of 2003 serves as a stark warning of the deadly impact the urban heat island effect can have on human mortality. That August, a persistent high-pressure system over Western Europe blocked Atlantic rainfall and funneled hot, dry air from North Africa into the region. The average summer temperature was about 3°C higher than the 1961–1990 mean—five standard deviations above the norm. France had already experienced an unusually warm and dry spring that year, and in August, Paris endured nine consecutive days above 35°C, peaking at 39°C on August 12. Nighttime temperatures rose even more sharply, winds were nearly calm, and relative humidity plummeted. High temperatures concentrated over downtown Paris at night and shifted to the industrial suburbs during the day.

The death toll was catastrophic, as more than 70,000 Parisians lost their lives. A study using National Oceanic and Atmospheric Administration (NOAA) satellite data examined the homes of 482 elderly individuals, half of whom died during the heatwave. Researchers developed daily and cumulative temperature indicators revaling that nighttime temperatures—not daytime highs—were the primary drivers of this mortality. A mere 0.5°C increase in nighttime temperature doubled the risk of death in this vulnerable population. The study concluded that increasing urban vegetation by just 2 to 3 percent could significantly reduce heat-related fatalities while also improving neighborhood livability.

Sea-Surface Temperatures

The temperature of the oceans is as essential to world climate as the temperature of land surfaces, if not more so. Just as satellites have transformed how we monitor land temperatures, they also play a crucial role in observing the oceans. Seventy-one percent of Earth's surface is covered by ocean water, and over 90 percent of solar energy reaching the surface is stored in the oceans, as demonstrated in Figure 13.6.

The sea surface is in constant exchange with the atmosphere, mediated by a complex web of physical, chemical, and biological processes. Among the

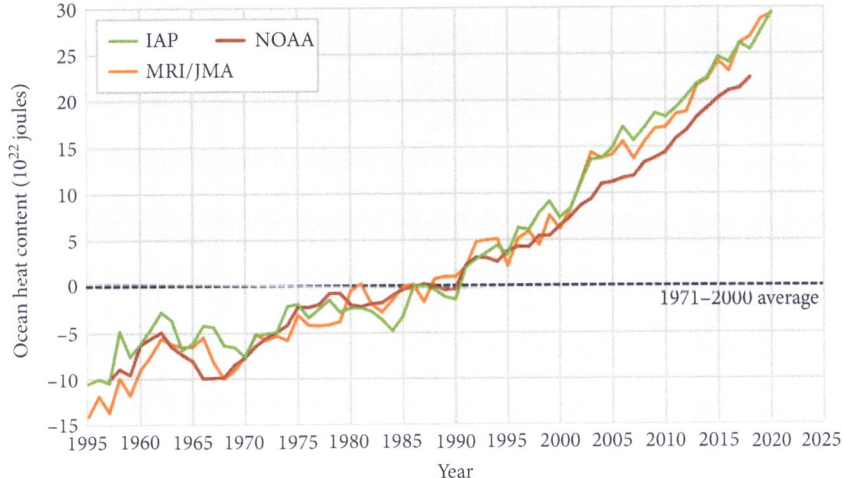

Figure 13.6 Change in heat content in the top 2,000 m of the world's oceans since 1955, relative to the 1971–2000 average. The three curves represent independent calculations made by NOAA, China's Institute of Atmospheric Physics (IAP), and Japan's Meteorological Research Institute. For perspective, an increase of 1×10^{22} joules in ocean heat content is not only about 17 times humanity's total annual energy use—it is also equivalent to the energy released by roughly 160 million Hiroshima bombs.
Environmental Protection Agency

most critical processes is evaporation, in which surface water transforms into vapor, rises into the atmosphere and is transported over vast distances before condensing and falling as precipitation. This mechanism plays a central role in the global redistribution of freshwater. The ocean also acts as a vast thermal reservoir, absorbing solar heat and releasing it slowly over time. This capacity to store and buffer heat helps moderate temperature extremes, regulating climate both locally and globally.

Scientists estimate that the oceans have absorbed approximately 25–30 percent of all CO_2 emitted since the Industrial Revolution.[1] When a CO_2 molecule sinks into the deep ocean, it can remain there for hundreds to thousands of years before returning to the atmosphere. This long storage time means that even if CO_2 emissions stopped today, global warming would continue for centuries—no one knows how long. In polar regions, sea-surface temperature plays a crucial role in climate feedback, as discussed above. Ice reflects most of the solar energy it receives, while open seawater absorbs much of it. As temperatures rise, melting ice exposes more water, increasing heat absorption and further accelerating ice loss. This self-reinforcing cycle amplifies warming, making polar regions particularly vulnerable to climate change.

The exchange of heat between the ocean and atmosphere plays a crucial role in driving the El Niño cycle, a phenomenon that significantly influences global weather patterns. During an El Niño event, which occurs every two to seven years, a pool of warm water in the western Pacific shifts eastward, disrupting atmospheric circulation and triggering extreme weather, including droughts and floods in various parts of the world.

Compared to traditional ship and buoy measurements, by continuously monitoring sea-surface temperatures in real time, satellites offer a more comprehensive and immediate perspective on El Niño's development and impact.

Argo and Beyond

Mariners and scientists have long recognized the importance of sea-surface temperature. In the early 1500s, Spanish explorer Ponce de León noted a powerful current flowing from west to east across the Atlantic, slowing westward voyages from Europe while hastening the return trip. In 1768, Benjamin Franklin, then deputy postmaster general for the North American colonies, became curious about why British mail packets took longer to cross the Atlantic than did American vessels traveling in the reverse direction. He consulted seasoned ship captains, including his cousin Timothy Folger, a Nantucket whaler, who described a strong, warm current flowing up the southeastern coast of North America and across the Atlantic. This current, later named the Gulf Stream—despite oceanographers finding that it carried little water from the Gulf of Mexico—became the focus of Franklin's investigations. During his transatlantic voyages, he and his colleagues used thermometers lowered over the side to identify the warm current and they even mapped its path.

Early sea-surface temperature measurements were made by hauling up a bucket of water and inserting a thermometer, with accuracy varying based on the bucket's material—canvas or wood—and other factors. By the 1960s, more consistent readings were obtained using thermometers placed in the engine room intakes of ocean-going vessels, providing a long-term, albeit sparse, dataset.

In the 1950s, anchored buoys began to be used to monitor sea-surface temperatures. Over time, the technology evolved, culminating in creation of the international Argo program in the early twenty-first century. Argo employs a global array of thousands of autonomous drifting floats that measure temperature, salinity, density, and ocean currents. These floats are equipped with a sophisticated buoyancy system that allows them to adjust their depth,

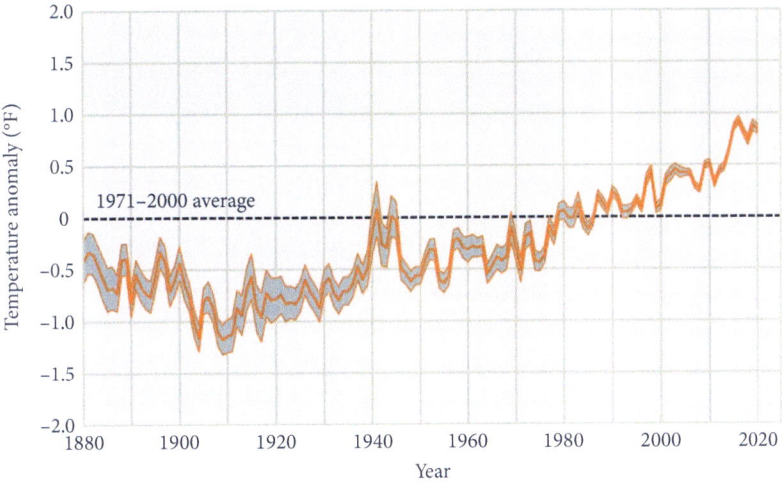

Figure 13.7 Average global sea-surface temperature, 1880–2020, NOAA

typically cycling between 2,000 meters and the surface. At regular intervals, each float rises through the water column, collecting data along the way. Upon reaching the surface, the float briefly extends an antenna above the waves to transmit data via satellite.

Argo's primary goal is not to measure sea-surface temperature per se, but to assess ocean heat content at intermediate depths—most notably around 1,000 m—where surface noise from waves, tides, and weather have little effect. This provides a far more stable and representative measure of long-term ocean warming, which is a key indicator of Earth's changing energy balance.

When viewed over time, not only have the deep oceans stored an almost incomprehensible amount of heat energy, as illustrated in Figure 13.6, but sea-surface temperatures have also increased, as shown in Figure 13.7. This chart provides independent evidence of the global warming observed in other charts and graphs. As with land-based impacts, these oceanic changes are monitored continuously by satellite, reinforcing their indispensable role in both diagnosis and mitigation of climate threats.

Hurricane Tracking

Satellites are indispensable tools for tracking hurricanes and forecasting their paths. They capture real-time imagery and measure critical variables such as wind speed, cloud structure, atmospheric moisture, and temperature—data essential for understanding storm dynamics. Geostationary satellites, like NOAA's Geostationary Operational Environmental Satellite (GOES) series, remain fixed over a specific region of Earth, delivering continuous, high-resolution observations that reveal the development and movement of hurricanes in near real time. Meanwhile, polar-orbiting satellites circle the globe from pole to pole, providing detailed, planet-wide coverage and vertical profiles of the atmosphere that are vital for improving forecast models. Working in tandem, these satellite systems enhance our ability to predict hurricanes more accurately, support early warning systems, and strengthen disaster preparedness efforts worldwide.

In 2005, Hurricane Katrina, shown in Figure 13.8, became the fourth-most intense hurricane to make landfall in the United States. Tracked by the GOES

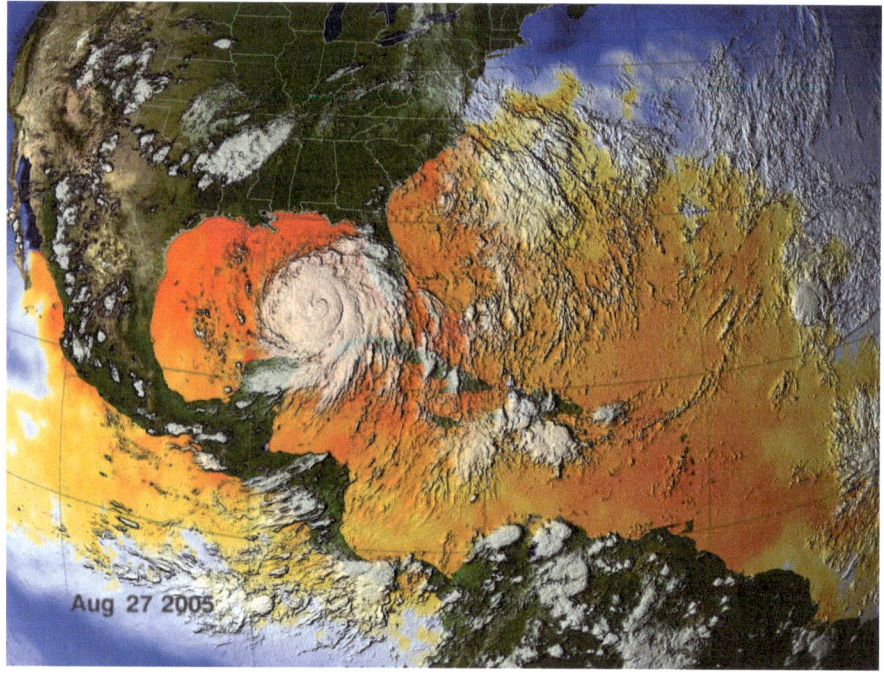

Sea Surface Temperature

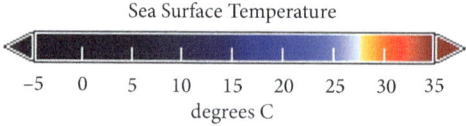

−5   0   5   10   15   20   25   30   35
degrees C

Figure 13.8 Sea-surface temperatures associated with Hurricane Katrina on August 27, 2005, measured by an instrument aboard the Aqua satellite. Most of the water in view is above the 26°C required to produce a hurricane.
NASA

satellites, Katrina initially struck the coast of southern Florida as a Category 1 storm. However, by the time it moved over the Gulf of Mexico, the warm seawater had intensified Katrina to a Category 5. Fortunately, the hurricane weakened to a Category 3 before making landfall again on the northern Gulf Coast, though it still caused extensive damage.

Ocean Acidification

When CO_2 dissolves in seawater, it forms carbonic acid, which impairs the ability of calcifying organisms such as corals, mollusks, and some plankton species, to produce and maintain their calcium carbonate skeletons and shells. Weakened skeletal structures make these organisms more vulnerable to predation and less able to compete for space and resources. For coral reefs, this results in reduced growth rates and less structural integrity, putting the entire reef ecosystem at risk. These "rainforests of the sea" host one of Earth's most diverse and valuable ecosystems. Corals are living animals that exist in a symbiotic relationship with a carbonate-secreting algae known as zooxanthellae, which provide corals with their color and nutrients through photosynthesis. Coral reefs shelter thousands of species, protect coastlines, and support regional economies through tourism. According to one estimate, the benefits provided by coral reefs amount to $10 billion annually. Sadly, coral and zooxanthellae are quite sensitive to the temperature of the surrounding seawater. When they die, they lose their distinctive colors, bleaching to a white, skeletal, disheartening remnant. Recognizing their importance, in the year 2000 NOAA established a program called Coral Reef Watch, which uses satellite observations of sea-surface temperatures to monitor reef health and predict bleaching.[2]

As long as global warming persists, coral reefs will continue to decline. Research indicates that 70–90 percent will be lost if the 1.5° limit of the Paris Agreement is exceeded—which already seems to have happened—and 99 percent will disappear under a 2° target, which is fast approaching.[3] Meanwhile, measurements of sea-surface temperature can at least predict areas where reefs are most at risk, enabling local organizations to mitigate the impact by closing reefs to divers and fishers, for example. However, ultimately, only one solution can save our precious coral reefs: immediate global action to combat global warming.[4]

PART VI

VEGETATION, WILDLIFE, AND OCEAN COLOR

14

Plants and Animals

The View from Space

Although satellite observations initially focused on inanimate objects, their application to monitoring the natural world began as early as the 1970s. In this chapter, we explore how satellites have been applied to assessing the health and abundance of plant and animal life—an application that the pioneers of satellite technology could scarcely have imagined.

The Normalized Difference Vegetation Index

While we may not often recognize it, our lives are deeply intertwined with and reliant on plants. They nourish us and the livestock we consume, and they offer shade, shelter, medicines, and construction materials. They absorb substantial amounts of carbon dioxide and replace it with life-giving oxygen—and they even clothe us.[1] Thus, the health of global vegetation is a vital element for humanity's future. Additionally, the destruction of rainforests, particularly in the Amazon and Indonesia, is already worsening global warming due to fewer plants surviving to absorb CO_2.

Launched in 1972, Landsat 1 was the first satellite specifically designed to image Earth's surface across multiple wavelengths. By capturing reflected sunlight in the visible and near-infrared spectrum, it enabled scientists to assess vegetation health and track changes in land cover. One of its earliest significant contributions was documenting the seasonal vegetation cycle of the American Great Plains, revealing patterns invisible from the ground. Landsat imagery showed the region's spring greening as crops and grasses flourished, followed by the expected browning in the fall as temperatures dropped and vegetation cover declined. These observations demonstrated the power of satellites for large-scale environmental monitoring, land management, and assessment of the vegetation-related impacts of global warming, such as drought stress, land-cover change, and changes in the length of the growing season. The success of Landsat 1 highlighted the importance of continuous Earth observation, laying the groundwork for the enduring Landsat program,

Sentinels in the Sky. James Lawrence Powell, Oxford University Press. © James Lawrence Powell (2026).
DOI: 10.1093/9780197842850.003.0014

which remains a vital tool for monitoring climate change, deforestation, and global land transformation.

This process, driven by chlorophyll, transforms carbon dioxide and water into energy-rich glucose and oxygen via a series of biochemical reactions. Photosynthesis underpins the food chain by converting solar energy into chemical energy stored in organic compounds, sustaining nearly all life on Earth and producing the oxygen essential for aerobic organisms.

Satellite monitoring of plant health relies on how vegetation interacts with sunlight across different parts of the electromagnetic spectrum. Chlorophyll—the primary pigment in photosynthesis—strongly absorbs visible light in the blue and red wavelengths while reflecting green, which is why healthy plants appear green to the human eye. Beyond the visible range, healthy vegetation also reflects a large portion of light in the near-infrared region, just beyond red. Though invisible to us, this near-infrared reflectance serves as a sensitive indicator of plant vitality: vigorous, photosynthetically active plants reflect more near-infrared light than those that are stressed, diseased, or dying. In contrast, vegetation suffering from drought, degradation, or other stressors tends to reflect more visible light and significantly less near-infrared radiation. This spectral shift provides a powerful diagnostic signal. By analyzing the ratio of reflected light in selected spectral bands, scientists compute vegetation indices that enable satellites to monitor plant health, productivity, and coverage with global scope and high temporal resolution.

By the early 1970s, satellite sensors could measure multiple wavelengths simultaneously. Later, satellite instruments such as MODIS (launched in 1999) expanded this capability with higher spatial and temporal resolution.[2] Consequently, those data contained information that, when correctly interpreted, could distinguish healthy from unhealthy vegetation automatically and over a wide area.

As is common in science, the key to ensuring the usability of data depends on presenting it in an easily understandable way. This prompted vegetation researchers to convert the amount of radiation reflected at visible and near-infrared wavelengths into a ratio known as the Normalized Difference Vegetation Index (NDVI). This metric reflects chlorophyll abundance and correlates strongly with plant health. The formula is:

$$NDVI = \frac{NIR - RED}{NIR + RED}$$

NIR represents light reflected in the near-infrared spectrum, while RED indicates light reflected in the red, visible range of the spectrum. Note that

NIR and RED are the only two pieces of data required to calculate the ratio. In the example shown in Figure 14.1, the healthy plant on the left reflects 50 percent of NIR but only 8 percent of RED, resulting in an NDVI of 0.72.

The ailing plant on the right in Figure 14.1 reflects less NIR and more RED, resulting in an NDVI of only 0.14. Thus, NDVI directly indicates plant health. Clouds, snowfields, water, and soil yield low or negative NDVI ratios and can easily be distinguished by satellites. NDVI has the advantage of being "normalized," ranging between −1 and +1, converting a multitude of complex data into a single, easily interpretable number.

NDVI has served as an index of plant health for several decades. It has now been complemented by various other indices, such as the Enhanced Vegetation Index, which adjusts the radiation data for atmospheric conditions and tree canopy background noise, proving particularly useful in areas with dense vegetation.

Satellite-based measurements of vegetation indices have become indispensable tools across a range of scientific disciplines. Among their most

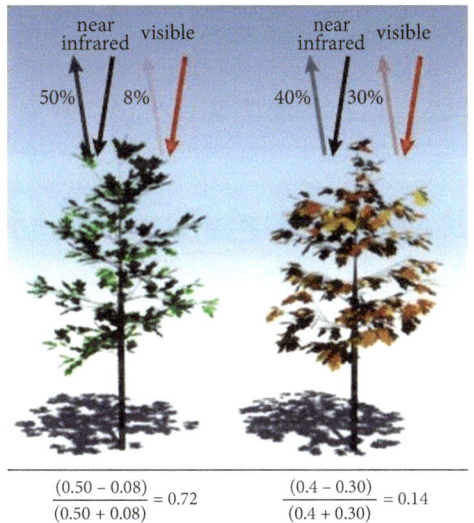

$$\frac{(0.50 - 0.08)}{(0.50 + 0.08)} = 0.72 \qquad \frac{(0.4 - 0.30)}{(0.4 + 0.30)} = 0.14$$

Figure 14.1 NDVI is calculated using the difference between near-infrared (NIR) and visible red light (RED) reflected by vegetation. Healthy vegetation (L) absorbs most of the visible red light while reflecting a large portion of NIR light. In contrast, unhealthy or sparse vegetation (R) reflects more visible RED and less NIR. NDVI values range from −1 to +1, with dense, healthy vegetation (such as a rainforest) having values close to +1, while barren surfaces like rocks, snow, or water typically have NDVI values near 0 or negative.

NASA Earth Observatory

important applications is the monitoring of global vegetation health in response to climate change. Since the early 1980s, NOAA has used NDVI to track global "greenness" on a weekly basis, providing a long-term record of plant vitality. Analysis of these data revealed that the growing season at northern latitudes lengthened beginning in the 1980s, a trend that has continued to the present day. NDVI thus offers robust evidence of climate-driven shifts in seasonal plant activity.[3]

Applications of NDVI

Vegetation indices have many uses in agriculture, including planning irrigation, optimizing fertilizer use, and timing harvesting. The index provides sufficient resolution to assess the health of individual plants in row crops, such as those in vineyards and orchards, even pinpointing locations where individual irrigation nozzles may be malfunctioning.

Insurers use NDVI to predict risk and quickly assess crop damage while processing claims resulting from hail, wind, frost, and drought. In the 1920s, long before the satellite age, Indian economist J. S. Chakravarti proposed using a technique known as index-based insurance to estimate property value and damage. Traditionally, insurance companies compensate policyholders for their actual losses, verified through on-site inspection and claims assessment. In contrast, Chakravarti's method determines payouts based on an index or metric rather than actual losses. These indices may include rainfall levels, temperature, earthquake magnitude, and, more recently, NDVI. Index-based insurance offers the advantage of rapid response across large areas at a significantly lower cost to insurers than on-the-ground assessments, particularly in regions where post-disaster damage evaluations are difficult or impossible. This approach has been tested in the United States, Canada, and several developing countries. For example, in Zimbabwe, insurance contracts that depend on NDVI assessments from satellites showed stronger correlations with yield losses than conventional rainfall indices and would be more affordable for smallholder farmers.[4] A similar study in Bihar, India, found comparable results, with even greater improvements when NDVI was used in conjunction with a rainfall index.[5] If this method can significantly lower premium prices, both insurers and policyholders may come to embrace it.

Vineyards are often divided into large blocks that may contain different varieties of grapes and exhibit different soil properties. Despite these differences, growers typically have had to view an entire block as the "minimum management unit" for cultivation and harvesting.[6] Beginning in the 1990s, by applying NDVI either from a commercial satellite company or NASA, vintners have been able to identify and monitor the variability within blocks. This advance allowed areas with lower NDVI values, likely caused by diseases and pests, to be targeted for special treatment.

Subsurface archaeological features such as walls, roads, and foundations influence the growth of vegetation above them. For instance, buried stone structures can change soil moisture and nutrient levels, resulting in variations in vegetation health and density. NDVI reveals areas of unusual vegetation growth that may lead to the discovery of archaeological sites, locate ancient Roman roads, identify leftover munitions, and plan urban green belts. Foresters apply NDVI to assess forest health, detect illegal logging, and manage resources by providing data on forest cover, biomass, and biodiversity.

Wildlife

One of the striking features of the Normalized Difference Vegetation Index (NDVI) is its wide-ranging utility in ecological research. As one might expect, NDVI has been widely used to assess the habitats and population dynamics of large herbivores, including African elephants, gazelles, red deer, reindeer, and wildebeest—species whose abundance and movement closely tracks the availability of vegetation. Yet scientists have found that NDVI, which measures "primary productivity"—the rate at which plants convert solar energy into biomass through photosynthesis—has implications that extend beyond herbivores. Because primary productivity underpins the entire food web, NDVI can also help explain the distribution and behavior of species such as birds, bears, lynx, and vervet monkeys, all of which tend to favor regions with high vegetation productivity and limited seasonal variability.

In the case of vervet monkeys, for example, researchers have found that the distances they travel each day closely correlate with monthly NDVI values—a reflection of shifting food availability. By linking vegetation dynamics to animal behavior, NDVI provides a powerful lens through which to view the ecological consequences of changing environments across a wide array of species.

Pests

NDVI, along with rainfall data, can also predict the size of mosquito populations and the subsequent spread of malaria. Several other diseases, such as schistosomiasis, Ebola, bubonic plague, leishmaniasis, and Rift Valley fever, have been found to correlate with NDVI.

The index is also used to track and monitor outbreaks of desert locusts. Throughout human history, these voracious creatures have been an indomitable scourge. In the list of the plagues inflicted by God upon the Egyptians, as described in the book of Exodus, locusts rank third, following the death of firstborn sons and three days of darkness. Today, the desert locust (*Schistocerca gregaria*) remains the most formidable migratory pest on Earth, notorious for its explosive reproductive capacity, long-distance mobility, and insatiable appetite for crops. Locust outbreaks are notoriously difficult to predict and control, often requiring coordinated ground and aerial pesticide campaigns across vast, remote regions. A major locust infestation that began in the Arabian Peninsula in 2018 illustrates the scale of the threat. Within three years, the swarms had spread across the Horn of Africa and reached as far as Tanzania and Iran, a spread of about 2,500 km, affecting millions of people and devastating croplands across several nations.

Yet even these resilient insects have a critical vulnerability: They flourish not in the driest deserts, but in moist soils with ample vegetation. Females preferentially lay eggs where recent rainfall has spurred plant growth, providing the developing young with the abundant food they need to survive and swarm. Scientists are using NDVI, in conjunction with soil moisture tracking, climate data, and wind direction, to forecast where locust outbreaks are likely to happen. They have begun applying machine learning and algorithms to enhance these predictions. The need for an enhanced response is urgent as climate change threatens to increase locust populations by causing more frequent desert rainstorms.

World Travelers

To conserve threatened and endangered species, conservationists must know as much as possible about every aspect of their lives: "where they eat, sleep, breathe, and mate."[7] Almost by definition, the more endangered a species, the harder it becomes to find and study in the wild. Historically, direct observation has often been limited to a single or a few individuals in a small area. However, many organisms travel in herds or flocks, and tracking has shown that some cover nearly inconceivable distances. One great white shark, tagged with electronic transmitters whose signals were retrieved by satellite, traveled from South Africa to Australia and back in nine months, covering a distance of more than 20,000 km.[8] A documented journey by a Leatherback Turtle, which would not seem to be the fleetest of marine species, made a 19,000-km round trip across the Pacific Ocean from Indonesia to the U.S. West Coast and back.[9]

The most remarkable long-distance traveler may be the bar-tailed godwit, shown in Figure 14.2. For years, Robert Gill, a biologist with the U.S. Geological Survey, has studied these birds, which were known to winter in Australia and New Zealand and summer in Alaska. Gill recruited a network of godwit spotters to keep track of the birds along their route, which, given the distance traveled, biologists had assumed must require a number of stopovers. Spotting the migrating birds was a hit-or-miss proposition, but a revolutionary change began in 2006 when satellite transmitters had become miniaturized sufficiently to allow them to be implanted in captured birds, who were then released and their journeys tracked. The result was mind-boggling: the godwits traversed the enormous central Pacific Ocean from Alaska to New Zealand, 11,680 km, in a single nonstop flight. In one documented case, the trip took only eight days without stops for food or rest.[10] As one researcher would later describe their journey: "It's not really like a marathon. It's more like a trip to the Moon."[11]

Figure 14.2 Bar-tailed Godwit, Global Traveler.
Wikimedia Commons

Marine Species

Few sights are as distressing as a massive leviathan—a humpback or sperm whale—stranded helplessly or lying dead on a beach. Most people encounter these tragic scenes only through photographs, as such strandings often occur in remote locations and may go undiscovered for weeks or even months—if they are found at all. The remoteness of many cetaceans from human settlements has hindered scientific study, resulting in a poor understanding of many populations. What is clear, however, is that whales and other marine mammals face increasing threats from human activities, including ocean pollution, ship strikes, entanglement, underwater noise, wildlife tourism, and climate change-driven disruptions to marine ecosystems.

The health of these populations matters because they act as "sentinel species," providing early warning signals of ecosystem health. Recognizing the urgency of their conservation, participants at the 2019 World Marine Mammal Conference in Barcelona endorsed goals aimed at improving responses to stranded marine mammals and protecting species at risk of extinction. Although networks have been established to monitor and assist stranded whales near populated coastlines, a more effective and timely detection system is critically needed to enhance intervention efforts and mitigate threats to these vital marine species.

Satellites have the advantage of imaging every square kilometer of the planet, including the most remote regions where strandings are likely to go undetected. However, manually reviewing images to detect whales is extremely time-consuming: scanning an area of 100 km^2 takes over three hours, which precludes real-time identification.[12] Scientists have started exploring automated analysis of very high resolution (VHR) images, which is a challenge due to the massive amounts of data that satellites record. Fortunately, thanks to advances in VHR technology, cloud computing, and machine learning, it appears that automated detection of whale strandings may soon become a reality. To that end, NOAA, Microsoft, and satellite company Maxar, along with other public- and private-sector partners, are developing an automated system called GeoAI (Geospatial Artificial Intelligence for Animals) for the automatic identification of marine mammals in satellite images.[13] A pilot program initially focuses on two species at high risk of extinction: the North Atlantic right whale and the much smaller Cook Inlet beluga whale, each of which is estimated to number only a few hundred individuals.

As VHR resolution has improved, scientists have shifted their focus to other marine species, including penguins and the spectacularly snouted elephant seals, shown in Figure 14.3. The Northern Elephant Seal, with its impressive males lounging contentedly among the assembled females, can be seen at

Figure 14.3 Female Northern Elephant Seals snuggled together for warmth. Near San Simeon, California, just south of Big Sur.
Wikimedia Commons

several locations along the California coast. However, to view the southern species, one must venture to remote sub-Antarctic and Antarctic waters, including South Georgia east of the Falklands (a brief stopover on some commercial cruises to Antarctica), Macquarie Island between New Zealand and Antarctica, and the Kerguelen and Crozet Archipelagos, French territories in the deep southern Indian Ocean. With the vast Antarctic ice cap looming beyond the southern horizon, these are the very waters most likely to be affected by global warming.

The significant size difference between the large males and smaller females allows for gender identification from satellite images, facilitating the monitoring of breeding opportunities. A study conducted by a team of French researchers focused on Kerguelen and Crozet, providing good news that is all too rare in these times of climate warming. Researchers have found that the total population of elephant seals on the two island groups—together accounting for roughly one-third of the global population—has increased significantly over the past five decades. On the Crozet Islands alone, the number of seals has more than tripled. This population growth appears to correlate with environmental changes associated with global warming, including the retreat of ice and the expansion of ice-free coastal areas that provide additional habitat suitable for breeding.[14] But rising sea levels driven by the same warming climate may ultimately counter or reverse this expansion by inundating low-lying shores, again reducing the area available for breeding.

Another group of polar specialists used VHR satellite imagery to estimate the population of Emperor penguins along the coastline of Antarctica, guided by the guano deposits left by penguins, which are visible as brown stains on the pristine white ice and snow, as shown in Figure 14.4.[15] The group counted the penguins at eleven sites and subsequently adjusted their estimates based on satellite imagery. This led to the discovery of four previously unknown colonies and confirmed the location of three suspected ones, increasing the total number to forty-six Emperor-penguin breeding colonies and an estimated population of nearly 600,000 adults. Further analysis using the Sentinel-2 satellite raised the number of colonies to sixty-one.[16] These results establish a baseline for future monitoring of the status of these iconic

Figure 14.4 An emperor penguin colony as revealed by the dark stains of their guano in a satellite image.

Maxar Technologies and Yale University

birds, which, like other polar species, are at risk as CO_2 emissions continue to rise.

The Internet of Animals

Argos (not to be confused with Argo), a collaborative initiative of the French Space Agency CNS (Centre national d'études spatiales), NOAA, and NASA, in the early 1980s became the first satellite program to track animal movements.[17] Transmitters were attached to captured animals, which were subsequently released. As the animals roamed, the transmitters sent signals that were detected by passing Argos satellites. By utilizing the Doppler shift recorded across multiple satellite passes, Argos triangulated the positions of individual animals within a few hundred meters. The raw location data were then transmitted via email to researchers on the ground. The early Argos transmitters were relatively large, often exceeding 1 kg, which made them suitable only for larger animals such as polar bears and King and Emperor penguins.

Argos demonstrated the potential of using satellites to monitor animal movements and set the stage for the benefits that miniaturization would bring, enabling the "Internet of Animals." This internet is somewhat analogous to the "Internet of Things" that some believe will soon emerge and allow our various gadgets to communicate.[18] Receivers have been miniaturized, and small solar panels eliminate the need for batteries, reducing weight and extending transmission time. Accelerometers continuously track animal movement and energy consumption over an animal's entire lifetime. Electronic devices can measure body temperature, heart rate, and even brain activity. These capabilities greatly increase the amount of data that needs to be processed, offering the benefits that "big data" has brought to other fields. These advances may enable attainment of the ultimate goal of remote animal studies: to predict where members of a species will move next.[19]

Several cutting-edge projects demonstrate the potential of the Internet of Animals. BirdCast uses AI to combine weather radar with live bird-tracking data to generate nightly migration forecasts. This enables cities to adjust artificial lighting, thereby reducing the number of birds that collide with buildings. Similarly, Whale Safe combines acoustic monitoring, satellite tracking, and ship movement data to provide real-time alerts that can prevent deadly whale–ship collisions along heavily trafficked coastlines.

Another initiative, Movebank, has compiled over 5 billion GNSS (Global Navigation Satellite System) tracking records of animals, allowing researchers to study migration patterns and habitat use.

Animal Migration

One challenge for researchers studying animal migration is the cost-benefit tradeoff of capturing and tagging individuals. This stressful process carries inherent risks and can potentially alter an animal's natural behavior. However, there are success stories: Captured in Germany in 1994 at the age of three, a female white stork named Princess was fitted with an Argos transmitter to track her migratory movements. Over her lifetime, Princess traveled approximately 10,000 km annually, once reaching a record flight speed of over 80 km/h. Unfortunately, because her transmitter relied on limited battery power, she had to be repeatedly recaptured and retagged. But still Princess survived until her death from natural causes in 2006.

A recent study utilizing solar-powered GNSS tags to track eleven populations of white storks uncovered a previously unrecorded stationary population in Uzbekistan, along with another group in Tunisia that makes multiple crossings of the Sahara Desert, a feat long deemed too demanding for these birds.

A study tracking three species of raptors initially used Argos satellite tags but transitioned to solar-powered GNSS devices in 2011. The researchers aimed to determine whether migration posed a higher mortality risk compared to periods when the birds were stationary. To identify deaths, they looked for two key indicators: a sudden loss of radio contact (if the transmitter had previously been working normally) or a lack of movement for several hours. When such evidence was detected, researchers conducted on-the-ground inspections to confirm whether the bird in question had died. The study revealed that mortality rates during migration were approximately six times higher than when the birds remained stationary. The most perilous period occurred when the raptors crossed the Sahara Desert on their northward journey to Europe, indicating that the harsh conditions and scarcity of resources in this region significantly heightened the risk of death.

Fishers are medium-sized, brown carnivores from the marten family, known for their rare ability to prey on porcupines. With few natural predators aside from humans, they were heavily trapped until the 1930s, when states began to impose restrictions. A key question in mammal conservation is how well species like the fisher can survive in landscapes increasingly dominated by humans. Using satellite imagery and GPS collar data, researchers found that fishers in suburban Albany, New York, consistently traveled along a limited number of forested corridors connecting fragmented woodlands. Satellite data helped map these corridors and identify how they thread through a matrix of roads, buildings, and residential developments. At one critical point, the corridor narrows sharply, creating a bottleneck where roads and structures converge—an area of heightened risk for human–wildlife conflict. If development were to block this passage, fishers would lose access to a significant portion of their habitat, surely causing local populations to decline.[20]

Another study focused on a large predator: the puma or cougar.[21] These big cats inhabit rugged terrain and require significant energy to locate and hunt prey. To study their movement and energy use, researchers developed a specialized GNSS collar equipped with accelerometers to track activity levels, movement patterns, and overall energy expenditure. Satellites recorded location data from collars placed on captured and released pumas, allowing scientists to analyze their hunting strategies.

The study identified two distinct hunting strategies among pumas: some use a low-energy, sit-and-wait approach, while others roam continuously, burning nearly twice as much energy. By outfitting the animals with accelerometer-equipped collars, researchers discovered that pumas conserve energy through stealth and precise movement, often remaining motionless until the moment of attack. When they do strike, their acceleration is finely tuned to the size of the prey, revealing why pumas—and most big cats—prefer ambush tactics over long pursuits: quick, calculated bursts minimize energy expenditure while maximizing success.

Given their iconic status and the fascination they inspire, wolves have been a prime subject for satellite tracking, especially following their reintroduction to Yellowstone National Park in the mid-1990s. A central question has been how the return of wolves would alter the behavior of their prey, particularly elk, and whether these changes would trigger a cascading effect, potentially influencing the health of the aspen and willow trees on which elk commonly feed. To investigate this issue, both elk and wolves have been geotagged and monitored. One study found that wolves had not altered elk behavior as much as anticipated. Despite the increased threat, the elk tended to return to the same wintering and summering areas year after year.[22]

In other research, scientists used satellite tracking to monitor the movement of bison to their wintering grounds at lower elevations outside Yellowstone. The data revealed how topography, habitat type, and roads influence bison behavior, helping to identify natural travel corridors and reveal the impact of human infrastructure on their movements. Grizzly bears in the park have also been subjects of satellite tracking; however, here ethical considerations come to the fore. These giants must be captured and tranquilized to fit or maintain the tracking collar, a procedure not for the faint of heart and certainly not beneficial for the bears.[23] In this case, the costs may outweigh the benefits.

ICARUS (the International Cooperation for Animal Research Using Space) is dedicated to tracking small flying animals using miniaturized transmitters. Established in 2002, the project made significant progress in 2018 when two Russian cosmonauts installed the ICARUS tracking system on the International Space Station (ISS). This high-altitude placement offers global coverage, enabling researchers to monitor animal movements over vast distances. The system operates using ultra high-frequency signals, which are transmitted from devices mounted on the animals and received by antennas on the International Space Station. Martin Wikelski, ICARUS project's director at the Max Planck Institute for Animal Behavior in Germany, likens it to a traffic app on a smartphone. Rather than monitoring a single car, the app compiles data from hundreds of vehicles to illustrate traffic patterns.[24]

ICARUS tracking devices, known as "bio-loggers," are remarkably compact—roughly the size of two fingernails—and weigh just 5 grams (see Figure 14.5). Consistent with the widely accepted guideline that tracking devices should not exceed 5 percent of an animal's body weight, they are well

Figure 14.5 An ovenbird fitted with a GNSS-enabled tracker.
Smithsonian

suited for small vertebrates such as songbirds, bats, and rodents. Although still too heavy for delicate species like monarch butterflies, researchers are developing even lighter technologies, including sensors that weigh as little as one gram, so as to track insect migrations. Data collected by these transmitters is sent to a receiver on the International Space Station, which relays it to ground stations. The resulting tracking information is made publicly available through online databases and can be accessed via smartphone apps such as *Animal Tracker*, allowing scientists and the public to follow animal movements in near real time.[23]

term. Sunlight plays a crucial role, but phytoplankton also rely on nutrients such as iron, nitrogen, and phosphorus, which rise from deep waters. Primary productivity governs the oceanic food chain, influencing everything from microbes to krill, fish, birds, and large marine mammals. It regulates the climate and supports biodiversity. Therefore, keeping track of changes in phytoplankton populations is essential.

Until recently, scientists believed that because chlorophyll levels undergo significant annual variation, at least thirty years of satellite-derived data would be necessary to differentiate the effects of climate change from the natural background. In a 2023 study, however, a team of researchers analyzed twenty-one years of satellite data and found this duration sufficient to demonstrate that nearly 60 percent of the world's ocean—an area larger than all landmasses combined—experienced a measurable color change over the period.[2] This shift is associated with warming surface waters, which generally lead to a decline in phytoplankton biomass and growth, ultimately diminishing ocean productivity.

Since phytoplankton are a major source of the greenish hue in ocean waters, their decline leads to a fainter green tint and a shift toward more transparent blue waters. This observed color change closely aligns with the 2019 predictions of MIT scientist Stephanie Dutkiewicz, who employed a computerized model to predict how climate change would affect ocean color as phytoplankton populations decreased.[3]

Algae and Red Tides

Microscopic phytoplankton belong to a broader group of photosynthetic organisms that also includes larger forms such as seaweed and kelp. When nutrients like phosphorus and nitrogen become abundant, these organisms can proliferate rapidly, resulting in dense blooms. Kelp forests offer a familiar example of macroscopic growth, while explosive increases in microscopic species can overwhelm smaller bodies of water. These blooms block sunlight, deplete oxygen, and sometimes release harmful toxins. Their ecological impact can be severe; for instance, such events are thought to have hindered ecosystem recovery after the great mass extinction at the end of the Paleozoic era.[4]

One of the most dangerous algal species is *Karenia brevis*, which can grow quickly and darken the sea in a phenomenon known as a red tide.[5] The toxins released by *K. brevis* have caused massive fish deaths in the Gulf of Mexico, poisoned shellfish, and even led to respiratory problems for nearby residents.

Since winds can move red tides, predicting their next location is important but challenging. However, by measuring chlorophyll concentration, which rises during a red tide, satellite data can help. NOAA maintains a Harmful Algal Bloom (HAB) Operational Forecast System in the Gulf of Mexico to identify potentially hazardous algal blooms.

Off the coast of Florida in mid-November 2004, images from the SeaWiFS instrument aboard the OrbView-2 satellite revealed a rise in chlorophyll concentration, possibly indicating an emerging red tide. In a few weeks, it had expanded to cover 1,000 km^2. Whether it was actually a red tide was uncertain, however, since a high chlorophyll concentration can also result from other causes, but there was a way to find out. Most of the sunlight that falls on algae is absorbed and converted to chemical energy through photosynthesis, but a small fraction is reflected at different wavelengths in a process known as fluorescence, which produces a faint red glow that a satellite radiometer can measure. Fluorescence measurements confirmed that an algal bloom caused by *K. brevis* was indeed occurring. The bloom led to fish kills and respiratory problems in humans on the nearby coasts as the toxin was converted into aerosols.

Another devastating bloom occurred at Paracas Bay, Peru, in April 2004, this one caused by the single-celled dinoflagellate *Gymnodinium sanguineum*. The satellite-based MODIS instrument was able to capture images on two of its radiation-detection bands, clearly showing the bloom. It led to the closing of a port and cost the Peruvian fishing industry an estimated $28.5 million in lost revenue.

Fisheries

Sea-surface temperature measurements help locate productive fishing areas, but they only detect the ocean's thin uppermost layer. Ocean color satellites, by contrast, penetrate several meters deep into the water column, revealing upwelling zones where cold, nutrient-rich waters rise from the depths. These upwelling areas, which temperature measurements might miss, create feeding hotspots that attract fish and drive their migration patterns. By detecting the chlorophyll and nutrients throughout the upper water column rather than just the surface skin, ocean color provides a more complete picture of where marine life congregates. India has developed an advanced satellite-based service that identifies promising fishing zones using color, which reflects chlorophyll concentration combined with sea-surface temperature. This information is crucial, as approximately 7 million people living on or near the Indian coastline rely on the fishing industry. By concentrating efforts where fish are most abundant, less productive areas are avoided, reducing overfishing pressure on depleted stocks and minimizing unnecessary fuel consumption. Improved catch efficiency lowers operational costs, with benefits that ripple across India's fishing economy.

Marine Animals

Ocean color has been used to track the behavior and migration patterns of various marine species, including whales, dolphins, seals, walruses, penguins, and sea turtles.[6] In one instance, scientists attached radio transmitters to loggerhead turtles that had been accidentally captured, then tagged and released, enabling the monitoring of their locations as they swam in search of jellyfish, barnacles, and crabs.[7] The turtles were observed to stay near the edge of the Transition Zone Chlorophyll Front (TZCF), which delineates the dynamic boundary in the Pacific Ocean between the cool, nutrient-rich waters of the subarctic and the warmer, nutrient-poor waters of the giant Pacific subtropical gyre. Consequently, the cooler, northern side of the front exhibits a higher concentration of chlorophyll compared to the warmer, southern side. The abundance of food, attributed to the elevated chlorophyll levels, makes the front a crucial feeding ground for many species. By observing ocean color, satellites can monitor the TZCF, where the mixing of water masses enhances phytoplankton abundance and marks the best feeding grounds.

Satellite data has also shown that fur seals from Kerguelen Island concentrate their foraging in areas with the highest chlorophyll concentrations.[8] Another study used sea-surface temperatures and chlorophyll levels to investigate the behavior of dolphin species near the Galápagos.[9] In one of the least expected applications of satellite observations, scientists analyzed the correlation between the locations of blue whale calls in the northwest Pacific and ocean color. They found that call locations correlated with high satellite-measured chlorophyll concentrations in spring, but not during other seasons when phytoplankton levels had declined.[10]

El Niño

The oceans are in a constant state of flux, with short-term variability that often obscures the signals of long-term climate change. Satellite observations are particularly well suited to distinguishing these patterns over time. In 1982, the onset of a major El Niño event provided an early test of the instrument's capabilities. CZCS data revealed a sharp decline in phytoplankton pigment concentrations near the Galápagos Islands, signaling a significant drop in primary productivity and making it harder for seabirds and marine mammals to reproduce.

One of the most intense El Niño events on record occurred in 1997–1998, coinciding with the launch of the SeaWiFS satellite, which began observing Earth on August 1, 1997. During this event, the warming of surface waters disrupted the normal upwelling of cold, nutrient-rich water, cutting off the supply of nutrients essential for phytoplankton growth. As a result, chlorophyll concentrations dropped markedly, signaling a sharp decline in phytoplankton abundance. This reduction had cascading effects on marine ecosystems and on fisheries, especially in regions that depend on nutrient-rich upwelling zones. More significantly, the decline in phytoplankton reduced the uptake of atmospheric CO_2. The loss of oceanic carbon absorption during this El Niño was substantial—roughly equivalent to half the annual fossil fuel CO_2 emissions of the United States at the time (about 3 billion metric tons). With less carbon drawn down by the ocean, more remained in the atmosphere, amplifying global warming.

At the same time, El Niño-induced drought across Southeast Asia, the Amazon, and Africa triggered extensive wildfires, releasing large amounts of CO_2 into the atmosphere and further intensifying its climatic impact. The following year, as sea-surface temperatures cooled, a La Niña event developed, strengthening upwelling and fueling one of the largest phytoplankton blooms ever recorded by satellites.

Bioluminescence

The ancient Romans and Greeks were aware that some sea creatures glow with an eerie internal light. Through the centuries, voyagers have noticed that sometimes at night, entire regions of the sea become luminous.[11] In the *Voyage of the Beagle*, Charles Darwin wrote:

> While sailing in these latitudes on one very dark night, the sea presented a wonderful and most beautiful spectacle. There was a fresh breeze, and every part of the surface, which during the day is seen as foam, now glowed with a pale light. The vessel drove before her bows two billows of liquid phosphorus, and in her wake she was followed by a milky train. As far as the eye reached, the crest of every wave was bright, and the sky above the horizon, from the reflected glare of these livid flames, was not so utterly obscure, as over the rest of the heavens.[12]

More than a century later, bioluminescence may have saved the life of astronaut James Lovell, who would go on to return the crippled Apollo 13 spacecraft safely. As John Edward Huth wrote in *The Lost Art of Finding Our Way*:

> In 1954 [Lovell] took off from the aircraft carrier USS Shangri-La in the Sea of Japan. His electronic navigation systems failed, and he had no way of determining where the Shangri-La was located. After switching off lights in his cabin, he noticed a glowing trail kicked up in the wake of the aircraft carrier. He then followed this trail of light churned by the propellors and landed.[13]

In Lovell's case, the bioluminescence appears to have been stirred up by the ship's propeller, but the illumination of large areas, such as the one that Darwin observed, has defied explanation to the present day. That said, in a 2021 article, a group of scientists described discovering a luminous area roughly the size of New England in the sea south of Java.[14] They studied satellite imagery from 2012 to 2021 and found a number of similar large-scale bioluminescence events, the smallest of which was still 100 times the size of Manhattan. These results stemmed from the use of a new sensor called the Day/Night Band (DNB), which is designed for low-light imaging. These findings did not solve the riddle of the milky seas, but they did show how satellite observations could help research vessels find these areas and study them close up.

Other Applications of Ocean Color

Observations of ocean color by satellite have many applications in addition to those we have reviewed.[15] A sample of other applications would include the following:

- Predict where endangered Right Whales will find food.
- Monitor remediation and recovery from oil spills.
- Combined with sea-surface temperatures, assist ship navigation and warn ships and oil drilling rigs of adverse conditions.
- Detect changes in sea-bottom topography.
- Monitor plumes where large, sediment-laden rivers like the Ganges and Mississippi enter the sea.
- Deliver water-clarity data to diving and de-mining operations in the Middle East.
- Predict the distribution of stinging jellyfish and dangerous sea nettles.
- Monitor high bacterial levels and phytoplankton blooms that can lead to cholera outbreaks.
- Identify and monitor hurricanes and typhoons.

PART VII
LOOKING OUTWARD

16

Telescopes in Space

Until now, we have examined satellites that observe Earth's land and oceans from above. Yet these same orbital platforms, when pointed outward, become powerful astronomical observatories. Positioned hundreds of kilometers above Earth's surface, space telescopes escape the atmospheric turbulence that blurs ground-based observations. Free from this interference, they achieve extraordinary clarity and can detect far fainter objects than even the largest terrestrial observatories. The impact of space-based astronomy extends beyond sharper images—these instruments have fundamentally transformed our understanding of the cosmos and humanity's place within it.

Lyman Spitzer's Vision

The idea of placing a telescope in orbit was first proposed by American astrophysicist Lyman Spitzer, who in 1946 wrote that "such a scientific tool, if practically feasible, could revolutionize astronomical techniques and open up completely new vistas of astronomical research."[1] He envisioned the equivalent of a 10-inch reflecting telescope at an altitude of about 800 km, placing it well above the distorting effects of Earth's atmosphere. This would allow, he wrote, "An object on Mars a mile in radius [to] be clearly recorded . . . while on the moon an object 50 feet across could be detected with visible radiation. This is at least ten times better than the typical performance of the best terrestrial telescopes." Spitzer emphasized that "[t]he chief contribution of such a radically new and more powerful instrument would be, not to supplement our present ideas of the universe we live in, but rather to uncover new phenomena not yet imagined, and perhaps to modify profoundly our basic concepts of space and time." Likely discoveries, he proposed, included the extent of the universe, the structure of galaxies and other clusters of stars, and the issue of whether life exists on Mars.

The far-sighted Spitzer wrote his article in 1945 and published it the following year—well before the Space Age had begun in earnest. At the time, Wernher von Braun's team in the United States and Sergei Korolev's team in

Sentinels in the Sky. James Lawrence Powell, Oxford University Press. © James Lawrence Powell (2026).
DOI: 10.1093/9780197842850.003.0016

the Soviet Union were only beginning to test captured German A-4 rockets. The very idea of launching even a small artificial satellite remained a distant and uncertain prospect, making Spitzer's foresight all the more remarkable.

As the Cold War deepened, the emphasis would soon shift to rockets designed to carry missiles rather than satellites. Spitzer was one of a handful of true visionaries in science, capable of conceiving a critical experiment long before it became technically feasible.

The first person to use a telescope for scientific observation was the renowned Italian "Father of Science," Galileo Galilei. In 1609, he built his own instrument—a tube about 1.7 m long with a convex objective lens at one end and a concave eyepiece at the other. It magnified objects roughly thirty times, allowing Galileo to see craters and mountains on the Moon (though no signs of life), observe the phases of Venus, and discover four moons orbiting Jupiter—findings that strongly supported Copernicus's heliocentric model of the solar system.

Galileo's device was a *refracting telescope*, so called because it gathered and bent light using lenses. Refracting telescopes are still favored by amateur astronomers, but they suffer a drawback: as light passes through a lens, it is dispersed into its component wavelengths, much like Newton's prism. Because these wavelengths bend by different amounts, they do not converge at a single point, resulting in *chromatic aberration*—a blurring of the image due to color fringing. To solve this problem, Newton devised a new type of telescope that used a mirror rather than a lens to gather and redirect light. Since mirrors reflect rather than refract, they avoid chromatic aberration entirely. Reflecting telescopes also offer a practical advantage: the light can be bounced from one mirror to another before reaching the eyepiece or detector.

Newton's contemporary, Frenchman Laurent Cassegrain, introduced a more complex instrument that improved performance and compactness. Both types are used today for most Earth-bound professional astronomy. As illustrated in Figure 16.1, light entering a Cassegrain telescope first reflects off a primary concave mirror, which focuses the light toward a secondary convex mirror. The secondary mirror then redirects the light back through a small hole in the center of the primary mirror, where it reaches the eyepiece or imaging sensor located at the rear of the telescope. This folded optical path allows the telescope to achieve a long effective focal length while maintaining a compact and manageable size, making it especially useful for astronomical observations.

When we position our eye at the focal point of a telescope, we perceive light in what is known as the "visible spectrum," which is the part of the electromagnetic spectrum that we can detect without assistance. As illustrated

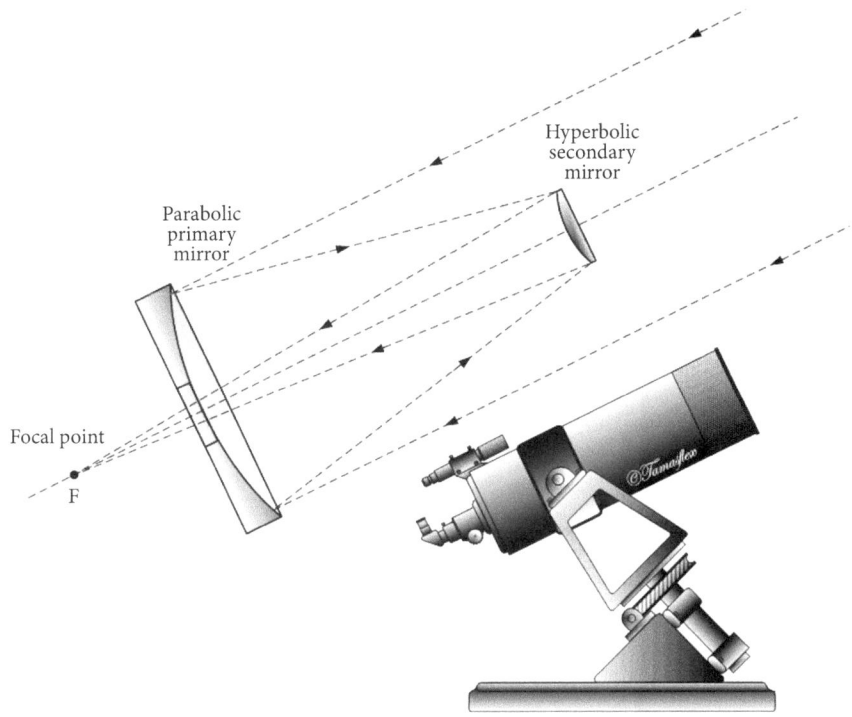

Figure 16.1 Diagram of a Cassegrain telescope. The light bounces off two mirrors and into the eyepiece at the focal point.

Wikipedia Commons. By Szőcs Tamás Tamasflex—Own work, CC BY-SA 3.0

in Figure 10.1, repeated here for convenience in Figure 16.2, the visible portion constitutes merely a small segment of the total range of electromagnetic waves. Author Andrew May provides a useful analogy to the eighty-eight keys on a piano.[2] If the key with the lowest note signifies radio waves with the lowest frequency, and the highest note signifies gamma rays with the highest frequency, a single piano key represents the entire visible region. However, by using electronic receivers instead of our own eyes, scientists detect the complete range of spectra arriving from space.

Radio Astronomy

The first electromagnetic waves detected outside the visible spectrum were radio waves transmitted through the ionosphere. Like many scientific discoveries, this one was serendipitous, made by young American physicist and engineer Karl Jansky, who worked at Bell Telephone Laboratories, which had

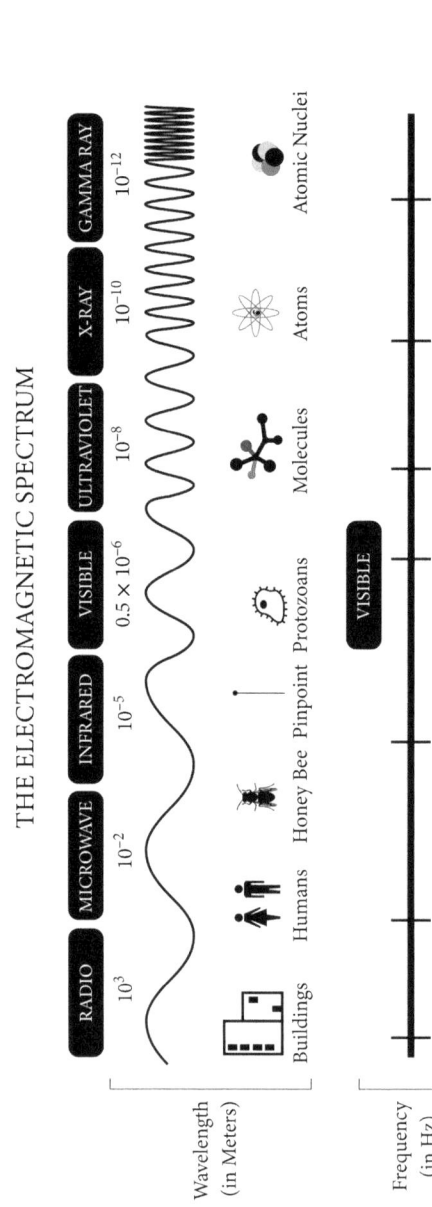

Figure 16.2 The electromagnetic spectrum.
Jonathan Urie, CC-BY-SA-3.0, via Wikimedia Commons

become a hub of innovation in mid-twentieth-century American science and engineering. Bell aimed to develop a service that would enable telephones to transmit and receive radio waves across the Atlantic Ocean, a concept with significant profit potential for the company. Assigned by Bell to determine whether static would interfere with the transmissions, Jansky built the world's first radio telescope, which, as shown in Figure 16.3, looked nothing like anyone's vision of a telescope or any other scientific instrument. Nevertheless, the ramshackle contraption worked well enough to allow Jansky to detect random static, which he believed originated from thunderstorms. However, he also heard a persistent hissing noise that he could barely distinguish from the background. At first, he thought the hiss must come from the Sun, but he was able to rule that out and conclude that instead it emanated from the center of the Milky Way, specifically from the constellation Sagittarius. Jansky presented his findings to a small audience at a radio engineers' convention,

Figure 16.3 Karl Jansky and his radio antenna in the early 1930s. This ungainly contraption consisted of an airplane wing mounted on automobile wheels in a potato field. As shown by the circular track and wheels, it could be rotated to point to different regions of the sky. It led to a revolution in astronomy.
Wikipedia Commons. Image courtesy of NRAO/AUI

only one of whom showed any interest. But sometimes, one is enough.[3] The interested party was John Kraus, who, after World War II, established a radio observatory at Ohio State University. Reflecting on Jansky's discovery, Kraus wrote, "In 1930 essentially all that we knew about the heavens had come from what we could see or photograph. Karl Jansky changed all that. A universe of radio sounds to which mankind had been deaf since time immemorial now suddenly burst forth in full chorus."[4]

As Andrew May sums up,

> Almost as many new astronomical phenomena were revealed through radio astronomy as had been discovered in three centuries of observations with optical telescopes. The invisible cold gas that permeates galaxies, and spreads out far beyond their previously known boundaries, suddenly became visible. Unimaginably powerful radio sources were discovered lurking at the centres of some galaxies, blasting out far more energy than can be accounted for by the visible stars there. Completely new objects, given fanciful names like pulsars and quasars, solved longstanding mysteries of stellar and galactic evolution – and raised many new mysteries of their own.[5]

Nancy Grace Roman

The successful detection of radio waves, one segment of the electromagnetic spectrum, highlighted the broader potential of observing other wavelengths. Many of these could be best studied—or in some cases, only studied—using a space-based telescope, like the one proposed by Lyman Spitzer.

To oversee astronomical research, which was to be a key program of the new NASA, the agency appointed an astronomer with an outstanding reputation. The unusual aspect of this appointment was that the new appointee was a woman, at a time when women were largely excluded from professional astronomy. This exclusion had been true for an unconscionably long period, not only in astronomy but in much of science.[6] The autobiography of Nancy Grace Roman, NASA's first Chief of Astronomy, is a must-read for anyone interested in the history of astronomy, especially for today's women scientists who, like her, may have been held back because of their gender.[7] Her story is shocking to read, even for someone who remembers those days from the perspective of another virtually all-male field, geology (Figure 16.4).

Nancy Grace Roman joined NASA in February 1959 with a mission to launch a space telescope. She knew it would prove revolutionary, and she worked tirelessly to bring it about. As her NASA biography, written in 2019, notes,

Figure 16.4 Nancy Grace Roman, the "Mother of Hubble."
NASA

Over the next two decades, Roman devoted herself to the cause of securing congressional approval and funding for what would become the Hubble Space Telescope. Roman led science and planning committees, brought together scientists and engineers, briefed executive branch officials, and lobbied NASA officials to approve a project as financially and scientifically ambitious as the Large Space Telescope, as it was then called. When talking to Congress, she often quipped that "for the price of one night at the movies each year, each American would receive 15 years of exciting discoveries." She was almost right—each American has now received over 30 years of discoveries!

From the location where it would be built to the size of its instruments, Roman was involved in every crucial decision about the telescope. Equipped with a strong scientific background and knowledge of the government's innerworkings, she [also] acted as the link between astronomers and Congress. . . . At NASA, Roman felt, people didn't look at her gender, but at the value of her work.[8]

Roman's first major success came with the launch of Orbiting Astronomical Observatory 2 in December 1968. The mission was groundbreaking in its focus on ultraviolet radiation from space—a region of the spectrum that had remained largely unexplored by astronomers due to Earth's atmospheric absorption. A decade later, she helped lead the development of the International Ultraviolet Explorer (IUE), a joint mission between NASA and the European Space Agency (ESA). Like many pioneering space observatories,

IUE was originally intended to operate for just three years, but it far exceeded expectations, remaining in service for eighteen.

Although the path for a woman astronomer during that era was long and challenging, Roman persevered and placed herself in the right place at the right time. Still, it was a difficult journey, as she recounted in an interview. When asked which mission in her career was most significant to her, Roman responded,

> The one I feel most proud of is the IUE. And the reason I say that is that I think every other major project that I was involved with would have occurred sooner or later without me. I don't think IUE would have. I carried it on almost single handedly. I was fighting people . . . who saw it as competition for X-ray funds.[9]

Roman's determined support led to more than 100,000 observations by the IUE, which have been used in more than 4,000 peer-reviewed articles.[10] These observations have contributed to many areas of astronomy, including how stars evolve, what fills the space between them, and the nature of energetic galaxies.

First Telescope in Space

NASA had been planning since the early 1960s to send a large telescope into space, guided by a committee led by Lyman Spitzer, who, as we have seen, originated the idea. One question was how the data obtained by the telescope could be retrieved. At that time, to return film from space, the U.S. military retrieved canisters dropped from orbiting satellites. Meanwhile, the Russians developed the film onboard their spacecraft, scanned data electronically and transmitted the digitized images back to Earth. Neither method would work for the envisioned space telescope, which, for one thing, would orbit at an altitude of about 500 km. By the 1970s, scientists had developed charge-coupled devices (CCDs) to capture images in space. Unlike photographic plates, which had to be returned and developed, CCDs turn light into electronic signals that can be processed instantly and sent back to Earth.

To reduce its size, the envisioned space telescope would employ a Cassegrain mirror to focus incoming electromagnetic waves. The planners envisioned a mirror 3 m in diameter; however, as shown by Tsiolkovsky's rocket equation, the larger the mirror, the heavier and more expensive the spacecraft needed to lift it into orbit. A primary goal for the planners was to create a space telescope capable of measuring the brightness of stars known as Cepheid variables, located in a cluster of galaxies called Virgo, 54 million light-years from Earth. Roman calculated that a smaller, 2.4 m mirror would still be capable of observing the Cepheid stars, thereby reducing the telescope's size and cost and allowing it to be launched from the payload bay of the Space Shuttle instead of from a rocket on Earth.[11]

For her many contributions, Roman has rightly been nicknamed "The Mother of Hubble": the name given to the new telescope. She was an early and vocal advocate for a space telescope, and she played a crucial role in securing the necessary funding, which was not an easy task at a time when NASA was focused on the Apollo program. Roman was deeply involved in the early stages of the Hubble Space Telescope's development, from its conceptual design to the planning of its scientific objectives. Perhaps even more importantly, she broke new ground for women in the field of astronomy and space science, illuminating a path for future generations.

Edwin Hubble

The first telescope in space was named for astronomer Edwin Hubble (see Figure 16.5). The story of this telescope begins in the 1920s when Hubble was working at the Mount Wilson Observatory near Los Angeles. Before his research, the Milky Way was believed to comprise the entire universe. Other observed galaxies were considered "nebulae" within the Milky Way. However, by observing the distant Cepheid variable stars, Hubble discovered that most were located far beyond the Milky Way. He realized that ours is just one galaxy among a multitude of galaxies, marking one of the most fundamental discoveries in the history of astronomy, if not in all of science.

Earlier, we discussed the crucial Doppler Effect, where the sound of a rapidly moving train, plane, or automobile traveling toward you rises in pitch. Then, as the object passes and moves away, the pitch decreases. Light waves experience the same effect, shifting toward lower infrared frequencies as the source recedes, in a phenomenon known as the "redshift." Although the Doppler shift of light is too subtle to notice in everyday life, the vast distances to stars make even small motions detectable. As a result, astronomers can use this shift to determine whether a star is moving toward Earth or is receding from it. Hubble not only used the perceived brightness of Cepheid stars to calculate their distance but also measured the frequency of light emitted from them, enabling him to use the Doppler Effect to determine their speed and direction.

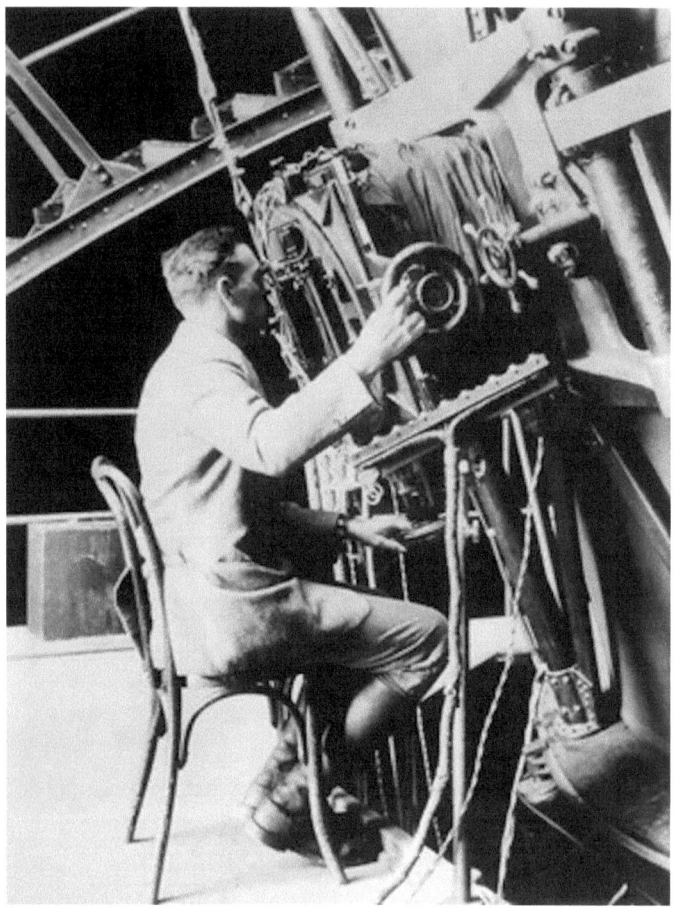

Figure 16.5 A young Edwin Hubble at Mount Wilson's 100-inch telescope, circa 1922.

Edwin Hubble Papers,/Courtesy of the Huntington Library

Astonishingly, Hubble discovered that the light from each of the galaxies he examined was redshifted, indicating that all were moving away from Earth. By measuring the magnitude of redshift, he could calculate how fast the galaxies were receding. Not only did Hubble find that all were moving away, but the more distant a galaxy, the faster it receded, which became known as Hubble's Law and established a foundational principle of modern cosmology. Hubble himself used a relatable analogy to illustrate how objects move in the expanding universe: a round loaf of raisin bread rising in an oven, as shown in Figure 16.6.

If every galaxy is moving away from ours, then the universe must be expanding. Even more remarkably, if one runs the tape backward, so to speak, imagining increasingly distant times in the past, the galaxies draw closer

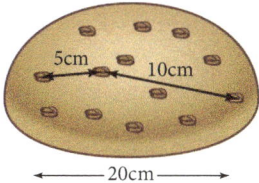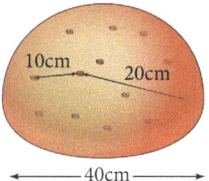

Figure 16.6 Analogy for Hubble's Law using raisins in a rising loaf of bread to represent galaxies. The farther apart two raisins are in the unheated loaf (left), the faster they move away from each other to give the new distances shown in the heated loaf (right.)
Wikimedia Commons

together until everything in the universe is compressed into a single point. This thought experiment established the foundation for the Big Bang theory, which envisions the universe beginning in a colossal explosion whose effects are still apparent as it continues to expand.

The Hubble Space Telescope

The new space telescope was scheduled to be launched by the Space Shuttle, shown in Figure 16.7. The reusable spacecraft had four components: (1) An Orbiter, which resembled an airplane and was designed to land on a runway upon its return. It housed the crew and payload. (2) Two Solid Rocket Boosters, which generated most of the thrust during launch. After using up their fuel, the boosters would detach from the main assembly and parachute back to Earth. (3) Three engines that controlled entry into orbit and were subsequently reused. (4) A large, orange fuel tank that supplied fuel to the main engines and, after separation, burned up upon reentering the atmosphere.

Since the Space Shuttle was going up anyway, it offered a cost-effective way to place satellites into orbit. Beyond affordability, launching from the Shuttle provided several other key advantages. It allowed satellites to be deployed at precise altitudes and trajectories, with the flexibility to make orbital adjustments as needed. More importantly, astronauts could conduct spacewalks to service the satellite, performing routine maintenance or addressing unexpected issues, which were bound to occur in the harsh environment of space. The Hubble Space Telescope, roughly the size of a bus, would have been difficult to launch using conventional rockets, but it fit comfortably within the Shuttle's payload bay.

Before the Hubble could be launched, however, a disaster occurred that no American who lived through it has forgotten. The *Challenger* shuttle had its maiden flight in April 1983, but on its tenth flight in January 1986, it

Figure 16.7 The Space Shuttle Challenger on the way to its last liftoff.
NASA

exploded seventy-three seconds after lift-off, killing all seven crew members. One was Christa McAuliffe, a high school social studies teacher from Concord, New Hampshire. She had been selected from more than 11,000 applicants to be the first civilian and teacher in space as part of NASA's Teacher in Space Project, a program intended to inspire students and promote public interest in science and space exploration. McAuliffe was scheduled to teach lessons from orbit during the mission. Her death deeply affected the nation.

Shuttle missions were suspended for three years while the cause of the accident was investigated. It turned out to be something so small and mundane that it was hard for people to grasp: rubber O-rings, designed to seal joints in the booster, had become too brittle on the unusually cold morning and had leaked hot pressurized gas that reached the external propellant tank, causing it to explode.

After the multiyear delay, on April 24, 1990, the Shuttle *Discovery* carried the Hubble telescope aloft and directly released it into orbit. Then trouble struck again. Due to faulty calibration of the equipment used to test Hubble's primary mirror, its shape was off by about 2,200 nm. (A nanometer is one billionth of a meter, or roughly one-fiftieth the thickness of a human hair.) This minuscule discrepancy was enough to cause light reflected off the edge of the mirror to focus slightly differently than light reflected off the center. The Hubble was so sensitive that this error was sufficient to blur images, undermining the justification for sending a telescope into space in the first place. Then the wisdom of making it repairable in space became evident. Just as when your car breaks down, you call a tow truck and mechanic, so in December 1993, the Shuttle *Endeavour* brought a crew of satellite repair technicians to the Space Station. Two pairs of astronauts performed five separate

Figure 16.8 The Pillars of Creation in the Eagle Nebula taken by the Hubble Space Telescope in 1995, showing masses of interstellar gas and dust. Stars are being created, while others are eroded by light from nearby stars.
NASA

spacewalks for a total of thirty-six hours. They installed equipment that corrected the infinitesimally misshapen mirror and restored the Hubble to its design specifications. Other astronauts would later service Hubble four more times, steadily upgrading its equipment.

The Hubble was placed into low earth orbit, where it completes a full revolution around the planet approximately every 95 minutes. Its orbit is inclined at 28.5°, a path that not only allows for optimal observations of the night sky but also ensures that servicing missions launched from the Space Shuttle remain feasible. To maintain its precise aim at celestial objects, Hubble relies on a system of internal flywheels and gyroscopes rather than fuel-based thrusters. These components allow for fine adjustments in orientation, ensuring that the telescope remains stable and can capture exceptionally detailed images of the universe, one of the most notable of which is shown in Figure 16.8. This image was far superior in every aspect to the best from a ground-based telescope. It did not suffer from atmospheric distortion, it possessed a much higher resolution, and it measured a broad range of radiation wavelengths.

The Hubble Constant

In his groundbreaking article, Lyman Spitzer predicted that a large space telescope would not only supplement the discoveries made by astronomers over the centuries but would also lead to findings that would profoundly reshape our fundamental concepts of space and time. He was more prescient than he could have anticipated, as the discoveries made by the Hubble telescope position it among the most significant scientific instruments in history. Indeed, it can be argued that by extending our knowledge back to the beginning of the universe, Hubble's breakthroughs represent the most fundamental of all.

The rate at which the universe is expanding is clearly one of the most fundamental physical factors. However, accurately measuring it has proven difficult. Edwin Hubble himself used a method based on his observations of the previously mentioned Cepheid variable stars. These stars possess the peculiar, yet invaluable, property of "pulsating": their temperature and brightness rise and fall in a regular pattern. In 1908, astronomer Henrietta Swan Leavitt discovered that the period of a Cepheid's repeating throb—how often it pulses—is directly related to its brightness. However, the brightness of a Cepheid star also depends on its distance from Earth. An equation illustrating these two facts would have brightness on both sides. By eliminating the common factor, brightness, the resulting equation demonstrated that by knowing the pulsation rate, astronomers could directly calculate the distance to the star. As noted, this method enabled Edwin Hubble to demonstrate that some Cepheid variable stars were so distant that they lay beyond the Milky Way, proving that they belonged to entirely separate galaxies. In doing so, he settled a pivotal debate over whether the Milky Way encompassed the universe or was one galaxy among many.

Hubble analyzed existing spectroscopic data to determine how fast galaxies were moving away from us, and then he graphed these recession velocities against the galaxies' distances. This analysis revealed a striking linear relationship: the farther away a galaxy is, the faster it recedes. The slope of this line became known as the Hubble Constant. Figure 16.9 shows a modern version of Hubble's original diagram, which is now based on far more precise measurements from the Hubble Space Telescope. These observations yield a Hubble Constant of 68 km per second per megaparsec—meaning that a galaxy located one megaparsec away (3.26 million light-years) recedes at 68 km/sec, while a galaxy twice as distant recedes twice as fast. This value closely matches recent measurements from the Planck mission, confirming that space itself is expanding uniformly throughout the universe.

Note that if the distance to a galaxy is measured in kilometers, the slope of the line in Figure 16.9 has distance in both the numerator and the denominator. Thus, as scientists say, "the units cancel," and what remains has the dimension of time.[12] This "Hubble Time" represents the duration a set of galaxies, all starting at the same time and at the speed derived from their redshifts, would need to reach their current separation distances. Working

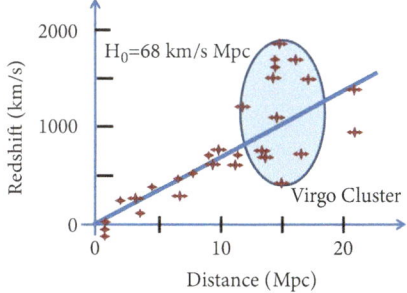

Figure 16.9 Hubble Diagram plotting distance in megaparsecs (1 megaparsec is about 3.26 million light-years or 3×10^{13} kilometers) vs. the amount of redshift in km/second. The data come from observation of the Cepheid stars.

Wikimedia Commons

backward, we find that the likely time of origin indicated by Figure 16.9—in other words, the age of the universe—is about 13.8 billion years. Compare this with the age of our solar system, determined through radiometric dating of meteorites, which is 4.5 billion years. The universe is three times as old as both Earth and the solar system.

This brings up an interesting scientific footnote. Einstein's original equations of general relativity did not allow for a static, unchanging universe—the prevailing assumption at the time—so he modified them by adding a term known as the *cosmological constant*. This additional term provided a repulsive force to counteract gravity and preserve the idea of a stable, nonexpanding cosmos. However, Hubble's discovery that the universe is expanding rendered the modification unnecessary, prompting Einstein to call the cosmological constant his "greatest mistake." In 1931, recognizing the importance of Hubble's findings, Einstein visited Mount Wilson Observatory to thank him personally. Ironically, modern cosmology has revived interest in the cosmological constant, which is now interpreted as a form of dark energy—an essential component in explaining the accelerating expansion of the universe.[13]

17
Pillars of Creation

Dark Matter and Dark Energy

Although the Hubble Space Telescope is best known for its breathtaking images, some of its most significant discoveries concern objects that cannot be seen or photographed. The notion that the universe harbors more matter than we can see dates back to the renowned physicist Lord Kelvin. In the 1880s, he analyzed the orbital speeds of stars circling the core of our galaxy to estimate its total gravitational pull. His calculations indicated that the galaxy contained far more matter than could be accounted for by its luminous stars alone. This discrepancy led Kelvin to propose that "[m]any of our stars, perhaps a great majority of them, may be dark bodies."[1] Forty years later, Swiss scientist Fritz Zwicky, working at the California Institute of Technology, studied the Coma cluster of galaxies and reached the same conclusion, naming the invisible substance "dunkle Materie," or dark matter.

The hypothesis did not gain broader acceptance until the 1970s, when American astronomer Vera Rubin provided compelling observational evidence. She found that stars near the outer edges of spiral galaxies orbit at nearly the same speed as those closer to the center. According to Newtonian dynamics, stars farther from the galactic core should orbit more slowly because the gravitational pull weakens with distance. The unexpected uniformity in orbital speeds suggested the presence of an unseen mass—dark matter—whose gravity alters the balance between inward pull and outward centrifugal acceleration and prevents these outward stars from accelerating. Calculations indicated that there must be five to ten times more dark matter than visible. Her contributions to astronomy have been honored by the Vera C. Rubin Observatory in Chile, which has begun to map the universe in unprecedented detail and is already producing revolutionary discoveries about transient phenomena, near-Earth asteroids, and the structure of our galaxy. With its 8.4-m mirror and the world's largest digital camera, the observatory has begun to photograph the entire visible southern sky every few nights, creating a dynamic, time-lapse record of the cosmos that promises

Sentinels in the Sky. James Lawrence Powell, Oxford University Press. © James Lawrence Powell (2026). DOI: 10.1093/9780197842850.003.0017

to reveal millions of previously unknown galaxies, track dangerous asteroids, and further illuminate the dark matter mysteries that Rubin herself pioneered.

Quasars (quasi-stellar bodies), the brightest objects in the universe, provide some of the strongest evidence for dark matter. When light from a distant quasar passes through a massive galaxy, it bends due to gravitational lensing—often by an amount that visible matter alone cannot account for—suggesting the presence of hidden mass. When the geometry is just right, the curvature of spacetime causes the light from a quasar to split into four different images, creating a sort of cosmological hall of mirrors, as shown in Figure 17.1.

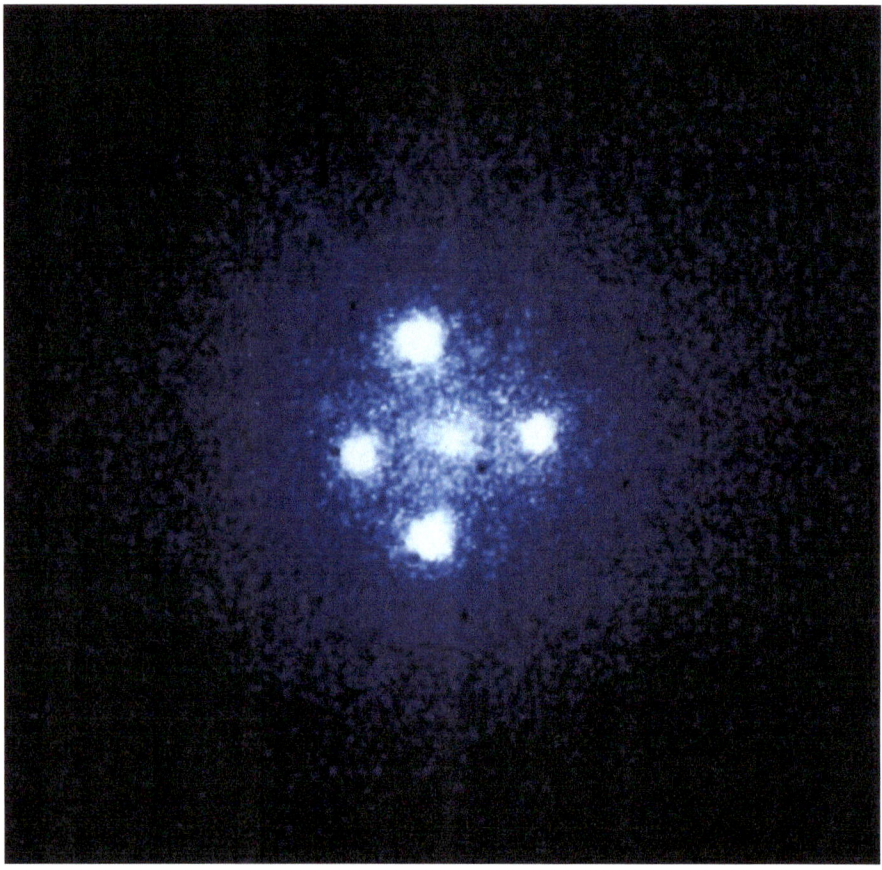

Figure 17.1 Einstein Cross: repeated images of a single distant quasar due to gravitational lensing caused by dark matter, providing further confirmation of Einstein's general relativity. The actual image is the one in the center. Photo taken by the European Space Agency's Faint Object Camera onboard the Hubble Space Telescope.
Wikimedia Commons

This image, known as the Einstein Cross, also serves as another confirmation of general relativity.

Further evidence comes from the gas surrounding quasars—extremely luminous, active galactic nuclei powered by supermassive black holes; the gas moves at speeds too great to be accounted for by the visible matter in their host galaxies alone. Even the host galaxies of quasars exhibit rotation patterns that suggest the gravitational influence of an unseen mass. Cosmological theory suggests that the earliest quasars, which formed when the universe was young, required massive dark matter halos to gather sufficient gas to ignite into stars.

Dark energy is an even more mysterious invisible component of the universe than dark matter. As noted, observations of distant supernovae suggest that the universe is not only expanding but is doing so at an accelerating rate. This indicates that some unknown force is counteracting gravity's effect, which would otherwise hold galaxies together and prevent expansion. Unlike dark matter, which is found in clusters and helps bind galaxies, dark energy appears to be uniformly distributed throughout space, suggesting that it is a property of the vacuum of space itself.

Deep Time

The Andromeda Galaxy—the nearest large galaxy to our Milky Way—lies 2.5 million light-years away. Thus, the light we now see from Andromeda began its journey toward us during the late Pliocene epoch, 2.5 million years ago. Thanks to the Hubble Space Telescope, we can now observe galaxies that are far more distant and ancient. One such example is GN-z11, located in the direction of the constellation Ursa Major. It lies 13.4 billion light-years from Earth, so we see it as it appeared when the universe was only a few hundred million years old. This cosmic dawn marks the era when the first stars and galaxies were taking shape. The ability to study such primordial galaxies provides astronomers with crucial insights into the origins and evolution of galaxies.

Exoplanets

Scientists have long believed, and science fiction writers have long imagined, that planets orbit other suns. However, even as late as the 1990s, no exoplanet had been discovered. Then in 1992, astronomers found two roughly

Earth-sized planets orbiting a pulsar—a rapidly spinning neutron star that emits precisely timed pulses of radiation.[2] In 2009, NASA launched the Kepler space telescope, specifically designed to detect Earth-like exoplanets and named after scientist Johannes Kepler (1571–1630), who formulated the three laws of planetary motion. NASA retired the Kepler Space Telescope in 2018 and, in the same year, launched the Transiting Exoplanet Survey Satellite (TESS), also known as Explorer 95. By April 7, 2025, Hubble, Kepler, TESS, and other missions had detected 7,443 planets in 5,105 planetary systems, 1,038 of which held multiple planets.[3] Strangely enough, many exoplanets are not tightly bound to a star, as are those in our system, but they wander through space until, perhaps, a star's gravity ensnares them.

In 2000, the Hubble Space Telescope became the first space mission to discover an exoplanet using the transit method (illustrated in Figure 17.2). When a planet moves, or "transits," across the face of a star, it blocks the star's brightness, so that seen from a distance the star appears to dim. Imagine an observer in a distant galaxy watching our solar system as Jupiter crosses in front of our Sun. The observer would see the Sun's brightness dim by about 1 percent, after which it would brighten again. This phenomenon would repeat indefinitely, evidence of an exoplanet orbiting its sun. In the case of Jupiter, a distant observer would see the dimming recur every 11.9 years, corresponding to its orbital period.[4] From this information, the extraterrestrial astronomer could calculate Jupiter's size, its distance from the Sun, and its density. For our much smaller Earth, the dimming observed from afar would amount to only about 0.01 percent, which might or might not be detectable. The distance of a planet from its star determines what astronomers call the

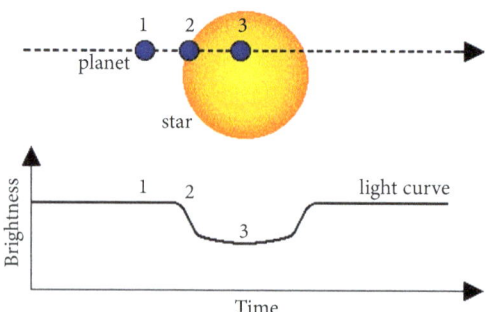

Figure 17.2 The light from a star dims as an orbiting planet passes in front of it.

European Space Agency

habitable, or "Goldilocks Zone," where liquid water could exist on a planet's surface and, therefore, where life as we know it might exist. In our solar system, the habitable zone is roughly between the orbits of Venus and Mars, with Earth positioned comfortably in the middle.

Not only could the Hubble detect the dimming effect of transiting exoplanets, but its spectrometers could recognize chemical fingerprints in the light passing through the planet's atmosphere and identify their composition.[5] This capability led to the discovery of methane—a potential biomarker often produced by biological processes on Earth—in the atmosphere of a planet located sixty-three light-years away.[6]

In 2018, Hubble observed the TRAPPIST-1 system, a compact group of seven planets orbiting a red dwarf star, with four planets lying within the habitable zone. These observations revealed that some of the planets lack the thick, hydrogen-rich atmospheres typical of gas giants like Neptune. Instead, their atmospheres may contain higher concentrations of molecules such as carbon dioxide, oxygen, and methane—compounds that are not only essential to life as we know it, but can also be produced by biological processes, making them tantalizing targets in the search for life beyond Earth.

It will come as no surprise that astronomers, having discovered a multitude of exoplanets, are now searching for exomoons. There must be countless more of them than exoplanets, as in our solar system, Jupiter has 95 moons and, when this writing began, Saturn had 146 moons, but in early 2025, scientists reported an additional 128, bringing the total to 274.[7] The transit method may be effective in detecting an exomoon if the orbits of an exoplanet and its moon are aligned just right, allowing the moon to cause detectable dimming of the star's light. A large moon can also induce a slight wobble in the orbits of both the planet and moon, which may be detectable.[8] Our Moon has a mass about 1.2 percent that of Earth, making it one of the largest moons relative to its planet in the solar system—though the Pluto–Charon system has an even higher mass ratio. Given the vast number of exoplanetary systems, it is likely that other "supermoons" exist beyond our solar system.

Observations by the Hubble Space Telescope have led to at least 15,000 peer-reviewed articles.[9] Newer missions have measured gamma rays and X-rays, infrared, ultraviolet, and microwave radiation, searched for exoplanets, counted stars, and much more.[10] Particularly noteworthy among Hubble's successors are the Cosmic Background Explorer (COBE) and the newest of all, the James Webb Space Telescope.

The Big Bang

What came to be known as the Big Bang theory of the cosmos originated in part from the work of a Belgian priest, astronomer, and professor of physics named Georges Lemaître, who earned a doctorate at MIT in 1927. In that same year, before Hubble's announcement, Lemaître published an article in which he deduced from the theory of general relativity that the universe must be expanding. Working backward in time, Lemaître proposed that the universe must have originated from a single "primeval atom." Einstein was not convinced, telling Lemaître, "From the point of view of Physics this seems to me abominable. What's the reason for such a brutal reaction?"[11] Then came Edwin Hubble's demonstration in 1929, based on the redshift of galaxies, that the universe actually is expanding. Many astronomers were unwilling to accept the nearly incomprehensible idea that the universe had once been a single point and came up with an alternative called the Steady State Theory. One of its proponents, British astronomer Fred Hoyle, had coined the phrase Big Bang, perhaps to disparage its rival, during an interview broadcast by the BBC in 1949. The term *Big Bang* went on to become one of the most recognizable in science.

The Steady State Theory asserts that matter is continuously created at just the right rate to maintain a constant density throughout space. There was no initial explosion; the cosmos had no beginning. It has always existed, timeless and infinite, maintaining a constant density as it expands. Admittedly, this concept is difficult for us to grasp, but no more so than the idea that the universe was once a single point.

Was there a way to distinguish between the two competing hypotheses of cosmic origins? In the 1940s, physicists Ralph Alpher and Robert Herman at the Applied Physics Laboratory of Johns Hopkins University proposed that if the universe had begun in a superheated explosion—the Big Bang—it would have left behind a pervasive background of microwave radiation. As the universe expanded and cooled from its initial hot, dense state, this radiation would have stretched, or redshifted, to longer wavelengths. Alpher and Herman calculated that this residual radiation should permeate the cosmos even today and would have cooled to a temperature of about—268°C. Such a temperature corresponds to that of a so-called blackbody—an ideal object that absorbs and re-emits all incident electromagnetic radiation with a specific spectral signature. Independently, Princeton physicist Robert Dicke reached a similar conclusion.

Two decades later, in 1964, Arno Penzias and Robert Wilson at Bell Labs in New Jersey were conducting radio astronomy experiments in support of satellite communications. Like Karl Jansky three decades earlier, they encountered an unexpected signal: a persistent, low-level background noise that appeared to come uniformly from all directions. At first, they suspected a mundane cause—pigeon droppings contaminating their antenna. But after thoroughly scrubbing the equipment, the signal remained unchanged. It became clear that the radiation was not of terrestrial or even galactic origin but cosmological, with a measured temperature that closely matched earlier theoretical predictions for the universe's background radiation.

Remarkably, just 40 km away, Robert Dicke and his research team at Princeton were preparing to search for this very signal—unaware that it had already been found. The two groups were operating in isolation, a striking example of the "silo effect" in scientific research. The detection of this microwave background radiation provided compelling evidence for the Big Bang model, effectively ruling out the competing steady-state theory. In recognition of their discovery, Penzias and Wilson received the 1978 Nobel Prize in Physics. Dicke, whose theoretical insight had anticipated the finding, was not recognized.

Further evidence for the expected "afterglow of the Big Bang," as it was termed, came from the Cosmic Background Explorer, a spacecraft specifically designed to study cosmic microwave radiation, which operated from 1989 to 1993. Data collected from its instruments showed that the intensity of the background radiation closely matched the predictions for the Big Bang across a wide range of frequencies, something the Steady State Theory could not begin to explain. The Big Bang theory has become the prevailing cosmological model, accounting not only for the origin of the cosmic microwave background but also for the large-scale structure of the universe and its observed expansion.

James Webb Space Telescope

The Hubble Space Telescope was capable of reaching much farther back in space and time than any telescope before it. However, by the 1990s, planning had already begun for a larger telescope that would explore even more distant and ancient objects and help answer a series of questions that Hubble's data could not address. In 2002, this new instrument was named the James Webb Space Telescope (JWST), in honor of the man who led NASA from 1961 to 1968 during the Mercury, Gemini, and Apollo programs. Naming the telescope after Webb marked a departure from the tradition of naming scientific satellites after astronomers and scientists, which began with Hubble and included figures like Planck, Herschel, Copernicus, and Kepler. Some came to regret this choice when, in 2015, a columnist reported that Webb, prior to joining NASA, had served as Undersecretary of State during a time when many gay men and lesbians were purged from the department as potential security risks. Four astronomers even published an article in *Scientific American* titled "The James Webb Telescope Needs to Be Renamed," but NASA ultimately decided against renaming it.[12]

The Hubble's mirror has a diameter of 2.4 m, while JWST's measures 6.5 m, resulting in a difference of 2.7 times. Since area scales with the square of the radius, however, the *area* of the JWST is 33 m^2 compared to Hubble's 4.5 m^2, giving it over seven times the collecting area, as illustrated in Figure 17.3. With increased size comes increased expense; in 2023 dollars, the JWST cost $10 billion, compared to $1.5 billion for the Hubble and $7 billion for another "Big Science" project, the atom-smashing Large Hadron Collider.[13] When the JWST launched, its cost amounted to 0.044 percent of U.S. GDP, slightly higher than the budget allocated to the National Science Foundation that year. However, it is important to remember that money spent on major science projects is used to pay wages and purchase goods; it does not disappear into some figurative black hole, never to be seen again, but recirculates within the larger economy.[14]

JWST was launched from the European Space Agency's Kourou spaceport in French Guiana to a location in space 1,500,000 km away, known as the Sun-Earth L2 Lagrange point. At a Lagrange point, of which there are five, the gravity of Earth and the Sun offset each other, allowing a spacecraft to remain there with minimal expenditure of fuel. The L2 point also provides a

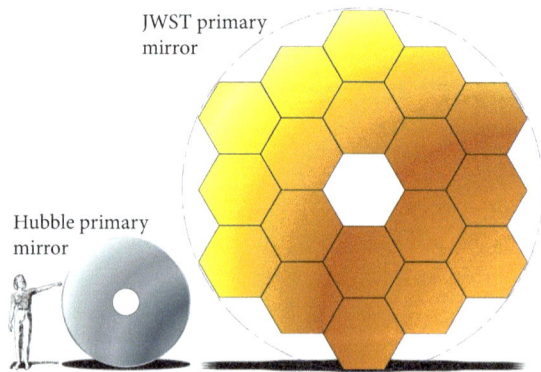

Figure 17.3 Area comparison of Hubble (4.5 sq. m) and JWST (33 sq. m).

Wikimedia Commons

near-constant temperature, suitable for a spacecraft's instruments, along with an unobstructed view of the entire sky. The downside is that L2 exposes a spacecraft to direct sunlight, which could blind it to much of the sky. Once in orbit, as shown in Figure 17.4, JWST deployed a large heat shield to block sunlight and protect its mirror and instruments.

JWST differs from Hubble in several important respects beyond its mirror size. For one, its sensors detect a much broader range of frequencies, including some in the visible spectrum but primarily in the infrared. There are several reasons for this. (1) As mentioned earlier, the visible region is a small fraction of the complete electromagnetic spectrum. Much of what interests astronomers lies outside the visible range. (2) By measuring the precise infrared wavelength of a distant object, scientists can determine its temperature, which changes as stars age. (3) Parts of a galaxy are often obscured by dust clouds that visible light cannot penetrate, whereas infrared radiation can "see" through such clouds. This capability is crucial in cosmology, allowing astronomers to observe phenomena associated with black holes, such as energetic jets and surrounding matter, whose light would otherwise be blocked.

The Webb telescope launched on Christmas Day, 2021 and returned its first images on July 11, 2022. Figure 17.5 illustrates the improvement in the

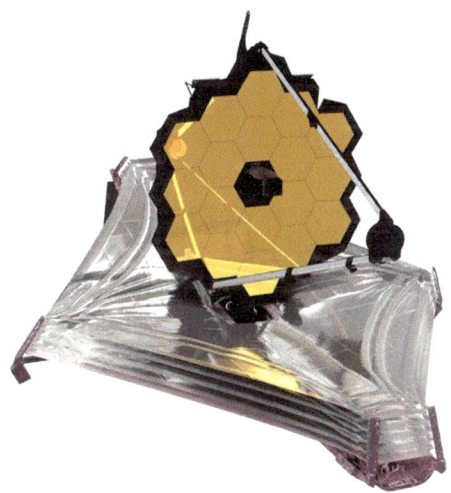

Figure 17.4 The James Webb Space Telescope showing the gold-plated beryllium mirror above and the sunshield below. The shield was intricately folded on launch, then unfolded remotely in space, something like doing origami with your eyes closed. NASA

Figure 17.5 The Pillars of Creation as imaged by the Hubble Space Telescope in 2014 (L) and by JWST (R), viewed in the near infrared.

NASA, ESA, CSA, STScI; J. DePasquale, A. Koekemoer, A. Pagan (STScI)

quality of JWST's images compared to those of Hubble, using the iconic Pillars of Creation image. The dramatic improvement stems from JWST's superior resolution in the infrared, revealing details previously hidden by dust.

One of the early objectives of the JWST was to refine measurements of the universe's rate of expansion, expressed by the Hubble Constant. Over the years, different methods for determining the Hubble Constant produced inconsistent results, fostering an ongoing scientific debate. Following the method pioneered by Edwin Hubble, JWST focused on ultrabright Cepheid variable stars in two distant galaxies. Compared to earlier observations made by the Hubble, JWST's data showed significantly less scatter, indicating a higher level of precision. However, despite this improvement, JWST's findings confirmed the same expansion rate as Hubble's earlier measurements.[15] This result suggests that the original Hubble-measured expansion rate was likely correct, reinforcing confidence in past observations.

Another application of JWST was directed at the galaxy GN-Z11, which was studied by the Hubble Telescope as noted above. It was found to be one of the youngest and most distant galaxies ever observed, forming only 430 million years after the universe itself.[16] The data supported the existence of a central "supermassive" black hole in GN-Z11, a region of spacetime where gravity is so strong that nothing—not even light—can escape its gravitational pull.

The JWST also detected a neutron star—the collapsed core of a massive star left behind at the center of supernova remnant SN 1987A, the closest observed supernova explosion since the invention of the telescope. In the extreme conditions of a supernova, protons and electrons are crushed together, fusing into neutrons, which gives neutron stars their name. First discovered in 1967 through radio pulses by British astronomers Jocelyn Bell Burnell and Antony Hewish, neutron stars are unimaginably dense, possess magnetic fields millions of times stronger than Earth's, and can rotate hundreds of times per second.

Just before the light from the 1987 supernova reached Earth, astronomers observed a 10-second burst of neutrinos: subatomic particles that are extremely light, neutral in charge, and interact weakly, if at all, with other matter, making them difficult to detect. Yet during the core collapse of a massive star, they are produced in large quantities and carry away a significant portion of the energy of the event. From the spectrographic data collected by JWST, scientists interpreted the neutrino burst as evidence that SN 1987A had left behind a neutron star, rather than a black hole, the alternate possibility.[17]

The JWST's focus on the infrared spectrum is particularly well suited for studying exoplanet atmospheres, as this is where the signatures of many key chemical markers of life are likely to be found.[18] Many exoplanets—most likely the majority—are small and at or beyond the limits of Hubble's detection. The increased resolution of the JWST will serve as a magnifier, enabling the discovery of many more of these distant worlds. The main instrument of the JWST, the Near-Infrared Spectrograph, achieves the seemingly impossible task of determining the temperature, mass, and chemical composition of 100 objects simultaneously. To accomplish this, it employs a "microshutter" to block the otherwise blinding light from an exoplanet's star.

Early results confirm JWST's potential for finding and analyzing exoplanets. One particularly interesting candidate is K2-18b, which is 8.6 times more massive than Earth, yet smaller than Neptune and located 120 light-years away, orbiting the cool dwarf star K2-18.[19] This exoplanet not only lies within the habitable zone of its star, but JWST has also detected in its spectrum an abundance of the biomarkers carbon dioxide and methane, making K2-18b a prime candidate for further study.

Lyman Spitzer was certainly right about the potential of space telescopes, as Hubble and JWST have transformed our understanding of the universe. Among Hubble's greatest discoveries was the evidence for the accelerating expansion of the cosmos, profoundly altering our view of dark matter and energy, as well as cosmic evolution. Its images revealed thousands of pre-viously unknown galaxies, shedding light on the early universe and galaxy formation. JWST, with its advanced infrared vision, has extended these find-ings by peering even further back in time, capturing the formation of the first galaxies and stars. It has also begun analyzing exoplanet atmospheres, detecting possible traces of water vapor and other key elements that may hint at the habitability of distant worlds. As this book is being written, JWST continues to produce new and stunning discoveries, often reported within days of observation. Readers who wish to stay current can visit NASA's JWST website.[20]

18

International Cooperation in Space

Early Competition

Satellites were born in the crucible of the Cold War, when the United States and the Soviet Union saw space as both a potential battlefield and a proving ground for technological dominance. The launch of Sputnik 1 in 1957 sent a shockwave through the Western world, signaling not just Soviet engineering prowess but the potential for surveillance, reconnaissance, and even warfare from orbit. The United States scrambled to respond, and on January 3, 1958, just four months later, Explorer 1 roared into space. The Space Race had begun.

In hindsight, the United States could likely have won the race to orbit even before it began, if only military and political leaders had unleashed Wernher von Braun and his team earlier. The Soviet Politburo had no such hesitation, and the USSR raced ahead with a string of historic firsts: the first animal in orbit (Laika), the first probe to escape Earth's gravity, the first crash landing on the Moon, the first photos of its hidden far side, the first man and later the first woman in space, the first spacecraft to orbit the Moon, and the first rover to trundle across another world. But for all these achievements, one absence loomed large—the Soviets never sent a cosmonaut to the Moon. Why?

The answer lies in a mixture of technology, politics, and economics. The Soviets lacked a powerful enough rocket, their space program was split by internal rivalries, and the death of Sergei Korolev in 1966 deprived them of their most brilliant leader. Unlike the United States, where John F. Kennedy and his successors threw their full support behind Apollo, the Soviet government wavered. The outlay for the Apollo program turned out to be staggering: in today's dollars, around $150 billion. The Soviet economy, already groaning under the weight of Cold War military spending, simply could not keep up.

By the late twentieth century, satellite technology had spread beyond the United States and the USSR. France entered the space race in 1965 with Astérix, followed by Japan's Ohsumi in 1970 and China's Dong Fang Hong 1 later that same year. China would go on to develop the BeiDou navigation system and the Gaofen fleet of high-resolution imaging satellites. India took a

Sentinels in the Sky. James Lawrence Powell, Oxford University Press. © James Lawrence Powell (2026).
DOI: 10.1093/9780197842850.003.0018

different approach, focusing on practical benefits. Since launching Aryabhata in 1975, India's satellite program has supported communications, weather forecasting, disaster response, and agriculture.

European nations pooled their resources to form the European Space Agency (ESA), which now oversees programs such as the Copernicus Earth observation satellites and the Galileo navigation network. Other nations soon followed. Japan's Himawari satellites track typhoons, while its Hayabusa probes pioneered asteroid sample-return missions. South Korea, initially reliant on foreign rockets, has since developed its own launch capabilities, sending its Arirang and KOMPSAT satellites into orbit. Brazil, the space leader of Latin America, focuses on monitoring the Amazon rainforest and partners with China on Earth observation satellites. The United Arab Emirates recently sent its Hope Probe to Mars and is developing its own high-resolution imaging satellites.

Today, more than seventy countries operate satellites, transforming what began as a two-nation competition into a global enterprise that serves humanity's shared needs—from tracking climate change and managing natural disasters to connecting remote communities and exploring the cosmos. This democratization of space technology demonstrates that the benefits of satellite observation, once the exclusive domain of superpowers, have become essential tools for nations of all sizes to address both local challenges and contribute to our collective understanding of Earth and beyond.

A New Era of Partnerships

The early years of space exploration were marked by rivalry, but collaboration has since become the rule. NASA and France's CNES have worked together on ocean-monitoring missions like TOPEX/Poseidon and Jason. ESA has partnered with NASA on projects ranging from the Hubble Space Telescope to the climate-focused Sentinel missions. India and France teamed up to study tropical weather, while NASA and the Indian Space Research Organisation (ISRO) launched the NASA-ISRO Synthetic Aperture Radar (NISAR) on July 30, 2025. This radar-imaging satellite will track environmental changes and natural disasters. Even China, often seen as a competitor rather than a collaborator, has joined forces with Brazil, France, Russia, and Pakistan on various space initiatives.

The greatest symbol of international cooperation in space is the International Space Station (ISS), shown in Figure 18.1. A marvel of engineering, the ISS orbits 400 km above Earth, a floating laboratory where astronauts

Figure 18.1 The International Space Station after undocking from the now-retired Endeavour space shuttle.

from rival nations conduct research together. Its origins trace back to the 1980s, when the United States planned a standalone space station called Freedom. Meanwhile, the USSR launched the modular station Mir in 1986. It hosted international guests, including NASA astronauts. As Cold War tensions thawed, Mir became a testbed for future collaboration. When it was decommissioned, the foundation was laid for what would become the ISS.

Construction of the ISS began in 1998 with the launch of the Russian Zarya module, followed by the American Unity module. Over the next decade, the station grew as nations contributed their own components: ESA added the Columbus laboratory, Japan installed its Kibo module, and Canada provided the robotic arm, Canadarm2. Today, five major space agencies—NASA, Roscosmos, European Space Agency, Japan Aerospace Exploration Agency, and Canadian Space Agency—operate the ISS. Astronauts conduct research on everything from the effects of microgravity on the human body to new materials and medical treatments.

The International Space Station also functions as a launch platform for small satellites, particularly CubeSats. As discussed in the next chapter, CubeSats are compact, modular satellites typically measuring 10 cm on each side and weighing about 1.3 kg per unit. They are deployed from the Japanese Kibo module, which is equipped with an airlock and robotic arms designed specifically for satellite release. Once released, these CubeSats enter the same low-Earth orbit as the ISS. While many remain in that orbit until atmospheric drag causes them to reenter and burn up, some are equipped with miniature propulsion systems that allow them to maneuver into slightly different orbits, extend their operational lifetimes, or optimize their observational vantage points.

The Future: Cooperation or Conflict?

For all its successes, the era of space collaboration faces an uncertain future. As geopolitical tensions rise, will the partnerships hold, or will space become yet another theater for competition? The warning signs are already visible. Rather than joining the ISS, China has pursued its own path by constructing the Tiangong space station, even as it engages in selective collaborations. Meanwhile, the establishment of the U.S. Space Force and similar military initiatives in Russia and China point to space being viewed increasingly through a strategic and defense-oriented lens. The development of anti-satellite weapons and cyberwarfare capabilities raises concerns about an arms race beyond Earth's atmosphere.

Then there is the matter of resources. The Moon's water ice and the mineral riches of asteroids may one day become flashpoints for territorial rivalry. While scientific exploration in space has historically thrived on international cooperation, the lure of economic profit and strategic dominance could just as easily pull nations apart. In the decades ahead, will space remain a realm of shared discovery, or will it come to reflect the same rivalries and divisions that so often shape international relations on Earth? How we answer that question may well determine the trajectory of space exploration for generations to come.

19

Golden Age or Gilded Age?

Proliferation

As of May 1, 2023, the Union of Concerned Scientists (UCS) estimated that around 7,560 satellites were in orbit.[1] Since then, due to the trend of launching multiple satellites simultaneously, the number is already considerably higher. It is estimated at 11,000 by *The Economist*, with requests for another one million on file with the International Telecommunication Union.[2]

Since the turn of the century, private corporations—not just national space programs—have been launching satellites into orbit. For instance, SpaceX has undertaken missions that were once the exclusive domain of NASA. These include transporting supplies and astronauts to the International Space Station (ISS), launching NASA science missions, and participating in the Artemis program aimed at sending astronauts into orbit around the Moon and subsequently to its surface. Other private companies such as Rocket Lab, Planet Labs, and Maxar are also deploying satellites for scientific purposes. Meanwhile, Blue Origin, founded by Amazon's Jeff Bezos, is constructing a commercial space station and has been selected by NASA to build a crew lander for the Artemis program. In February 2024, Houston-based Intuitive Machines successfully landed its Odysseus spacecraft near the Moon's South Pole, marking the first U.S. touchdown on the lunar surface in over fifty years. This trend shows no sign of slowing.

As the number of satellites in orbit increases, controlling them becomes more complex and costly. In the future, satellites will need to be autonomous, capable of sensing each other and adjusting automatically and safely. They must be able to inspect and service themselves and change their orbits on their own to avoid collisions while continuing their missions. Previously, satellites tended to send data to ground stations for analysis, but with miniaturization and increased computing power, satellites themselves will handle data processing and decision-making. As satellites are sent on more distant missions—where communication with Earth becomes delayed or unreliable—the demand for autonomy will only grow. Artificial intelligence, machine learning, and robotics will play an increasingly critical role.

Sentinels in the Sky. James Lawrence Powell, Oxford University Press. © James Lawrence Powell (2026).
DOI: 10.1093/9780197842850.003.0019

Small Satellites and Constellations

As rockets and payloads have increased in size, some satellites have become smaller, allowing for multiple launches that have significantly boosted satellite populations. The most common among these satellites are CubeSats, illustrated in Figure 19.1. When larger instruments are needed, multiple CubeSats—typically up to six—can be combined into a single payload.[3]

The CubeSat concept emerged in the late 1990s as part of a program at California Polytechnic State University (Cal Poly) in San Luis Obispo, California, designed to teach graduate students how to design and test satellites. The first CubeSats, which included designs from Cal Poly and various other institutions, were launched aboard a Russian Eurocket from the Plesetsk Cosmodrome in northern Russia. As of January 1, 2024, both academic

Figure 19.1 NCube-2, a Norwegian CubeSat. These devices measure 10 cm on each side.

Wikipedia Commons

and commercial organizations had launched over 2,300 CubeSats.[4] They have allowed many countries to enter the Space Age at a significantly lower cost, thus further democratizing space. In comparison to large, traditional satellites, CubeSats provide several additional advantages:

- Because they are standardized and use commercial off-the-shelf components, CubeSats can be developed and built much faster than larger satellites. This allows them to adopt technological improvements and miniaturization quickly.
- They cost far less to design, build, and launch. CubeSats often tag along on larger missions, such as those ferrying cargo and astronauts to the ISS.
- Their simplicity makes them ideal for educational institutions, helping to train graduates about satellite technology. Even a high school and an elementary school have successfully built and launched CubeSats.
- The lower cost and faster development time of CubeSats make them easily replaceable, reducing the cost of failure, promoting the testing of new technologies, and enabling startups and nongovernmental organizations to initiate space activities. For instance, in 2012, researchers published a review analyzing CubeSats and determining the scientific applications for which they would be most suitable. The authors categorized the applications into feasible, infeasible, and "problematic," with the latter indicating that CubeSats might not yield data of sufficient quality. Out of the fourteen technologies listed, only four were considered feasible for CubeSats. When this study was repeated in 2019, just seven years later, thirteen were found to be feasible and only one was deemed problematic.[5]

CubeSats may need to alter their orbits and orientations, but they are too small to carry sufficient rocket fuel for these adjustments. This necessitates finding alternative propulsion methods, some of which may later be adapted for larger satellites. Among those methods being tested are ion thrusters, where atoms of an inert gas like xenon receive an electric charge, allowing them to be expelled from the thruster, propelling the CubeSat in the opposite direction, akin to Tsiolkovsky's stranded boatman, introduced in Chapter 2. Another technology is a "solar sail," which can be deployed from a CubeSat and propelled by the pressure of sunlight. This method requires no fuel—only exposure to sunlight. From 2019 to 2022, the Planetary Society successfully used a solar sail to propel a CubeSat, as illustrated in Figure 19.2.[6]

Figure 19.2 Artist's conception of a solar sail attached to a CubeSat.

Solar sails may prove to be a breakthrough, further lowering the cost of operating CubeSats.

Small satellites are being used today not only for many of the applications we have reviewed so far, but also for surprising new ones. The Mars Cube One mission consisted of two CubeSats that traveled alongside NASA's InSight lander to Mars, marking the first time CubeSats had ventured into deep space. These small satellites proved that they could survive and function beyond Earth's orbit, successfully relaying real-time data from InSight as it descended to the surface of Mars. This feat demonstrated the potential for CubeSats to support interplanetary missions, providing communication, navigation, and reconnaissance capabilities in deep space. In the Colorado Ultraviolet Transit Experiment (CUTE) mission launched in September 2021, six CubeSats deployed a new ultraviolet spectrometer used to study exoplanet atmospheres. One of the most innovative projects was the use of a cluster of CubeSats to simulate a much larger antenna to take measurements from exoplanets.[7]

Low Earth Orbit

CubeSats and other small satellites benefit from deployment in low Earth orbit (LEO), typically at altitudes between 160 and 2,000 km—far below the 35,786-km altitude of geostationary satellites. Reaching LEO requires substantially less energy than higher orbits, dramatically reducing both manufacturing and launch costs. The International Space Station, orbiting between 330 and 435 km with its 51-degree inclination, serves as an ideal deployment platform for CubeSats that arrive as tag-along payloads. This orbit provides comprehensive Earth coverage with frequent overflights, perfect for small satellite operations.

LEO offers crucial advantages for modern applications. Signal latency—the round-trip time between satellite and ground—ranges from just 20 to 40 milliseconds, enabling real-time communications that can compete with terrestrial broadband networks. Ground stations for LEO satellites require less sophisticated equipment than those tracking distant spacecraft, further reducing operational costs. Additionally, satellites in these lower orbits naturally deorbit more quickly at mission's end, helping to mitigate the growing problem of space debris as atmospheric drag pulls them to burn up during reentry.

The strategic value of LEO satellite internet became starkly apparent following Russia's invasion of Ukraine in February 2022. When conventional infrastructure and internet services were destroyed, portable Starlink terminals paired with generators rapidly restored connectivity for both civilians and military forces, even in active combat zones. This demonstrated a fundamental reality of modern warfare: forces without reliable internet connectivity operate at a severe tactical disadvantage, making satellite communications a critical strategic asset accessible from anywhere on Earth.

Debris

History shows that most technological advances eventually bring accompanying problems. Airplanes revolutionized travel but were later used in World War II for carpet bombing and to deliver atomic bombs. Drones, a novelty only a few years ago, are now widely used in agriculture, environmental monitoring and conservation, search and rescue, surveying and mapping, and as one of the most lethal weapons in the arsenals of warring nations, as seen in the Russia–Ukraine War, where they have redefined military combat. Satellites are no exception; they also face a growing downside that threatens their future use.

As the number of satellites rises, the risk of collisions with one another and with manned spacecraft increases. Mission planners must already account for debris avoidance, which can shrink launch windows and require additional fuel for in-orbit maneuvers, thereby reducing payload capacity. Even small pieces of debris traveling at high velocities carry a tremendous punch of energy that can cause catastrophic damage to satellites, potentially rendering them inoperable or even destroying them. A 2-kg CubeSat traveling at orbital velocity carries roughly 60 MJ of kinetic energy—comparable to a standard sedan driving at highway speed. The International Space Station already has to maneuver to mitigate the risk of collision with space debris. A fatal impact would not need to destroy a crewed spacecraft—only puncture its hull, exposing those inside to the vacuum and lethal cold of space.

This proliferation of satellites introduces a frightening scenario called the Kessler Syndrome. First described by NASA scientist Donald J. Kessler in 1978, this phenomenon occurs when orbital debris reaches a critical density where collisions become self-perpetuating. Similar to a nuclear chain reaction—where each splitting atom releases neutrons that split more atoms in an escalating cascade—each satellite collision generates fragments that strike other satellites, creating more debris in an exponential progression. As destroyed satellites multiply into thousands of high-velocity projectiles, the probability of additional impacts rises dramatically. At its most catastrophic, this runaway cascade could transform LEO into an impenetrable barrier of speeding debris, making space operations impossible for decades or centuries and effectively trapping humanity on Earth beneath a shell of our own orbital wreckage.

To tackle the growing issue of space debris, satellite robots are being developed to clean up orbits by removing defunct satellites and space junk. These robots would capture or nudge objects out of their current paths and into Earth's atmosphere, where they would burn up upon reentry. As satellites in LEO near the end of their lives, operators lower their orbits, ensuring they reenter the atmosphere within 25 years, as recommended by various international guidelines, where they disintegrate due to atmospheric friction. Satellites in geostationary orbit, which are too high for atmospheric reentry, are often moved to a "graveyard orbit" several hundred kilometers above to free up space for operational satellites. However, these practices depend on careful planning and compliance and are not enforced globally.[8]

Another concern stems from the exotic chemistry of particles ablated from satellites during reentry. These particles often include metals such as aluminum, copper, lithium, and niobium, which are introduced through use of specialized alloys and thermal shielding materials in spacecraft construction. Once vaporized in the upper atmosphere, these elements may persist as fine particulates or react with atmospheric gases in ways that are not yet fully understood.

Copper, for example, acts as a catalyst, meaning it can accelerate chemical reactions without being consumed in the process. This property makes it particularly influential despite its trace concentrations. In the mesosphere and lower thermosphere, catalytic reactions involving metals shed by satellites could alter the delicate balance of trace gases. These might include ozone, nitric oxides, and hydroxyl radicals, all of which play critical roles in regulating atmospheric chemistry and radiative balance.

The problem is compounded by the lack of direct observational data and the complex dynamics of high-altitude atmospheric layers, where extreme temperatures, sparse molecular densities, and unusual ion–molecule interactions prevail. No one can yet say with certainty what new compounds may form or what longer-term effects might emerge from the cumulative presence of these exotic materials. As the number of satellite launches and reentries continues to grow—particularly with the expansion of megaconstellations—this question becomes more urgent, demanding detailed study through atmospheric modeling, laboratory simulations, and direct measurement campaigns.

Legal and regulatory complexities pose significant obstacles to efforts aimed at cleaning up space. The 1967 Outer Space Treaty stipulates that ownership of satellites and other orbital objects remains with the original launching state or operator—even after the object becomes defunct. As a result, regardless of the risk it may present, removing orbital debris requires prior consent from the owner. The 1972 Liability Convention, which governs responsibility for damage caused by space objects, also holds the launching nation—or the one authorizing a launch—liable for any damage caused by the space object, whether on Earth's surface or to aircraft in flight. As debris-removal technologies evolve, both governments and private entities must grapple with these legal and policy constraints to enable coordinated, effective cleanup efforts. This underscores the urgent need for updated international agreements to manage both active satellites and the growing threat of space debris.

We do not have to wait for the events envisioned by the Kessler Syndrome; they are already upon us. In 2007, China used a ground-based, medium-range ballistic missile to shoot down one of its aging weather satellites.[9] This destruction reportedly produced over 3,000 pieces of trackable debris, along with an estimated 150,000 particles smaller than 1 cm. China's test drew widespread international condemnation, particularly from the United States, Japan, and India, as well as from organizations concerned with the peaceful use of outer space. In June 2024, a decommissioned Russian satellite in low Earth orbit blew apart. The details remain unknown, but the *New York Times* noted that at the time of the event, the satellite passed over a Russian missile site.[10] India and the United States have also conducted antimissile exercises, despite the United States' previous statement that it would not engage in such tests. More antimissile tests seem inevitable.

Surveillance

For high-altitude spy flights over potentially hostile nations, satellites have replaced airplanes, as the following example illustrates. On May 1, 1960, at 70,000 ft above Sverdlovsk in Soviet airspace, American pilot Francis Gary Powers was flying a U-2 spy plane equipped with advanced photographic technology when a Soviet surface-to-air missile shot down his aircraft. Although the United States believed the U-2 spy program to be secret, Soviet authorities had been tracking the plane. The shootdown triggered a significant Cold War crisis, with the United States claiming that the U-2 was merely a weather research aircraft that had strayed off course after the pilot experienced oxygen difficulties. The explanation held only briefly. The Soviets soon presented Powers—who had miraculously survived—along with recovered photographic equipment and images of Soviet military sites. As a mark of the times, Americans were stunned that senior officials—including the president—had misled the public so openly. As a result, the planned summit between President Dwight D. Eisenhower and Soviet Premier Nikita Khrushchev was canceled, postponing discussions on arms control and other vital issues.

Nearly three decades later, on April 28, 1986, the Soviet nuclear reactor at Chernobyl, Ukraine, exploded. This time, it was the Soviets who lied, downplaying the incident and putting their own citizens at risk. The first indication that a serious nuclear incident had occurred came when workers at the Forsmark Nuclear Power Plant in Sweden, 1,100 km from Chernobyl, were found by routine radiation sensors to have arrived for work at the plant with radioactive particles on their clothing. Wind direction suggested that a nuclear incident had taken place in the western Soviet Union, but no disaster had been announced. By 1986, however, the satellite age had made it far more difficult to keep secrets from the global public. Three days after the explosion, U.S. Landsat 5 flew over Chernobyl. At an altitude of about 650 km, it could resolve objects on the ground and see what was happening inside the Chernobyl power plant, as shown in Figure 19.3. Landsat became the first civilian satellite to detect the disaster when its infrared radiometer revealed a bright red hotspot inside the plant—evidence that the graphite control rods were

CHERNOBYL   APRIL 22, 1986

CHERNOBYL NUCLEAR REACTOR SITE

April 22, 1986

0   1   2
KILOMETERS

N

Landsat-5 Thematic Mapper Bands 7,4,2 (RGB)

Figure 19.3 Chernobyl site from Landsat 5, April 22, 1986.
USGS

burning at high temperature. This indicated that the reactor's control systems had failed, likely allowing oxygen to enter the core leading to the release of radioactive material.

The Chernobyl disaster underscores the dual nature of modern satellite observation, which not only exposed Soviet attempts to conceal the explosion but also played a vital role in managing the response. Satellite imagery has made it nearly impossible for nations to hide certain activities, whether involving the movement of troops, military equipment, and aircraft during wartime; the locations of nuclear facilities, concentration camps, mining operations, and illegal shipping routes, or presenting evidence of war crimes, human rights abuses, and so on.

At the same time, satellite imagery plays a vital role in peaceful and humanitarian efforts. It is used to monitor environmental changes, track wildfires, manage agricultural resources, guide urban development, assess water quality, coordinate disaster response, and support public health initiatives. While the ability to observe and document events from orbit can be unsettling, it has also become an essential tool for tackling global challenges and ensuring transparency in ways that benefit society. Nations must find ways to maximize the benefits of satellite observation while minimizing its potential drawbacks.

Image Resolution

Image resolution—the size of the smallest object that can be clearly distinguished—has steadily improved since the 1980s. In response to privacy concerns, NOAA initially prohibited commercial satellites from selling imagery with resolution finer than 50 cm. In response to complaints from companies selling satellite images, in 2014 NOAA reduced the limit to 31 cm. In 2019, the Indian Space Research Organization, which is not bound by NOAA's regulations, launched its Cartosat-3 satellite, which has a black-and-white resolution of 25 cm. The satellite industry is advocating for a further reduction of the limit to 10 cm, roughly the size of your longest finger and technically feasible. NOAA took a step in that direction in August 2023 by lowering the limit to 16 cm and may well decrease it further. Meanwhile, nothing prevents countries like China and India from selling imagery at the lowest resolution they can achieve.

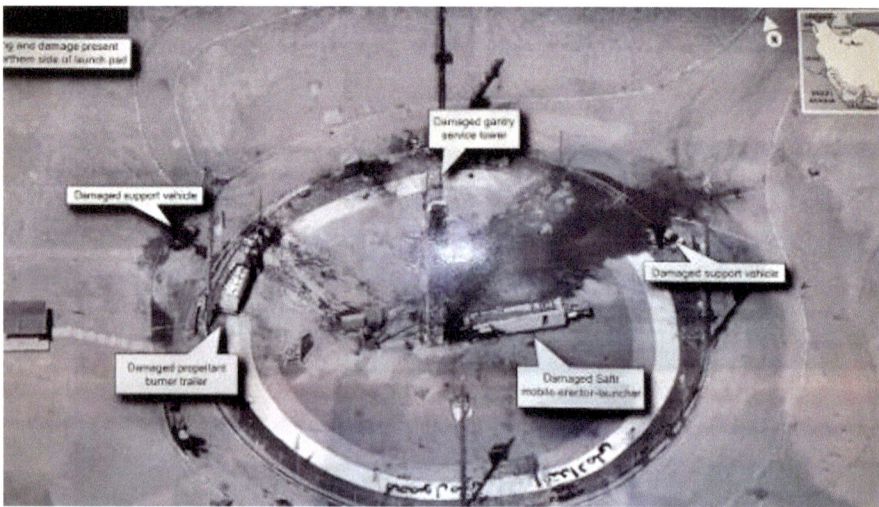

Figure 19.4　Heavily damaged launchpad in Iran, photographed by a U.S. satellite and tweeted by President Trump. Resolution: 10 cm.

Spaceflight Now

As an illustration of the futility of attempting to limit resolution, consider an incident in 2019, when President Donald Trump posted to his 80 million followers a satellite surveillance photo of an Iranian missile site in the aftermath of an explosion. Based on the lighting, angle, and estimated time of the photo, experts deduced that USA 224, an optical reconnaissance satellite, took the highly classified photograph, shown in Figure 19.4. Most striking was the image's sharpness—estimated at 10 cm or less in resolution.

Space Wars

In today's wars, space is emerging as the new battleground. A notable example can be seen in Ukraine, which, as noted, lost most of its satellite connections shortly after Russia's invasion in February 2022. Among these were services provided by Viasat, a U.S.-based satellite communications company that had been Ukraine's most popular provider. The shutdown was attributed to a Russian cyberattack, demonstrating that satellites can be rendered useless without being physically damaged. Modern military operations depend on satellites for the digital communications and real-time data exchange that coordinate forces across battlefields. Increasingly, without that ability they are blind.

One issue that has occupied the minds of political and military leaders, as well as the public, since Sputnik is the possibility that a satellite could carry and deploy a nuclear weapon. The concern in the 1950s was that a satellite might release a bomb that detonated near the ground, akin to those dropped on Hiroshima and Nagasaki. By the 1980s, under President Ronald Reagan, concern had shifted to the possibility of a nuclear detonation in space—one that could disable entire fleets of satellites, including those of the attacking nation.

Such an explosion would not only damage military and civilian communications but would also potentially cripple the satellites that provide various services we have reviewed, ranging from weather forecasting to internet service. A 2010 study found that astronauts—such as those aboard the International Space Station—would face a near certainty of death following a nuclear explosion in space.

Privacy

The challenge posed by the satellite age for policymakers and society, as exemplified by the Chernobyl incident, is how to balance the benefits of transparency with concerns about privacy, sovereignty, and security.

Privacy has become an increasingly significant issue. The combination of high-resolution imagery, big data, and facial recognition means that it is now technically feasible to identify individuals from satellite imagery. This is precisely the goal of Albedo Space, a Denver-based American startup.[11] The company is developing satellites that will operate at very low altitudes and capture images with 10-cm resolution, enabling the facial recognition of individual persons. If not Albedo Space, then another firm will undertake this task, for if it can be done, it will be.

It is not hard to imagine the dangerous effects of facial recognition. For example, consider the use of satellite imagery and facial recognition technology to identify in seconds voters waiting in line at polling places. So-called volunteer monitors could cross-reference these images with government or social media databases to identify individuals' political affiliations and voting histories—information that could be used to intimidate selected voters. Even the mere possibility of such surveillance could discourage many from voting in person, leading them to opt for mail-in ballots regardless of their preferences, or to sit out the election. In authoritarian regimes, facial recognition could be weaponized to target dissidents, protesters, and activists. Governments could use the data to harass, detain, or blacklist individuals, undermining freedoms of speech and assembly. The technology could be used to recognize whistleblowers and journalists, who could be identified and tracked when meeting confidential sources.

Insurance companies might implement satellite-based facial recognition to monitor risky behaviors of their policyholders, potentially denying coverage to individuals engaged in activities like extreme sports or smoking. Employers could carry out unauthorized surveillance on employees, monitoring attendance at job sites or tracking movements outside of work hours. If high-resolution facial recognition data were to become affordable for private individuals, it could empower stalkers, criminals, and abusive ex-partners to locate and track their victims with unprecedented precision, leaving them nowhere to escape. In such a world, privacy would be a relic of the past—and Big Brother a reality.

Treaties

More than 100 nations signed the 1967 Outer Space Treaty, but it did not address individual privacy rights, a concern that had not yet emerged. In 1986, the principles of the Outer Space Treaty were adopted and expanded in a United Nations statement on Remote Sensing Principles. Its primary message was that remote sensing from space should be conducted for the benefit of all countries, regardless of their level of economic or scientific development. Then, in 2018, the European Union implemented the General Data Protection Regulation, which aimed to protect privacy in an increasingly digital world. The regulation was designed to cover individuals' rights regarding data such as location, medical records, and social media activity. But does an identifiable satellite image of a person in a public crowd constitute personal data? And when disaster strikes, does the right to privacy outweigh the benefits of facial recognition?

Past agreements, though limited in scope, demonstrate the possibility of establishing global or regional frameworks to address the legal and privacy challenges posed by satellite surveillance. The need for such solutions is increasingly urgent as technology advances at a pace far outstripping the development of laws and international agreements. The principle of open skies—that no country can claim sovereignty over orbital space above the 100 km altitude that marks the beginning of space—must not come at the expense of individual privacy—nor should it legitimize orbital surveillance by an all-seeing authority. Safeguarding privacy in the satellite age will require deliberate efforts, sustained dialogue, and a commitment from nations to cooperate in setting meaningful regulations.

Priorities in Space: Science, Spectacle, or Both?

Of all government-supported research programs created since World War II, satellite observations have had the broadest reach and have led to some of the most transformative scientific discoveries. Yet can we be confident that the momentum behind satellite-based scientific research will continue? Nations and private companies will no doubt launch satellites for military, commercial, and operational purposes. But missions devoted purely to science—often without immediate commercial payoff—rely on sustained government and philanthropic investment. As space agencies increasingly prioritize high-profile crewed missions to the Moon and, eventually, Mars, they may divert resources from less spectacular but no less essential scientific endeavors. In the United States, rising skepticism toward science—amplified by opposition to climate research and the politicization of the pandemic response—threatens to push science funding, including for space-based research, further down the list of national priorities. Whether future satellite missions will receive the investment they merit remains an open question. The answer will shape not only the future of space science but also our understanding of Earth—and of humanity's place in the cosmos.

Postscript: Dark Skies Ahead

The predictions in the above paragraph were written in mid-Winter 2024–2025. Only six months into the second presidential term of Donald Trump, the predictions, and worse, have already started to come true.

The Trump administration has proposed unprecedented cuts to NASA's budget for fiscal year 2026, seeking to reduce the agency's funding by 24 percent—from $24.8 billion to $18.8 billion—which would represent the largest single-year cut in NASA's history and, adjusted for inflation, return the agency to 1961 funding levels. The cuts disproportionately target scientific research, with a devastating 47 percent reduction to the science budget that would terminate more than forty active missions, including fully functional spacecraft currently exploring the solar system. Among the casualties would be the Juno mission orbiting Jupiter, New Horizons exploring beyond Pluto, multiple Mars orbiters including Odyssey and MAVEN, the Mars Sample Return project, and Earth observation satellites that monitor carbon dioxide and climate change. Even the flagship Hubble and James Webb space telescopes would see operational budget cuts, while others like the Chandra X-ray Observatory and Fermi Gamma-ray Space Telescope would be shut down entirely.

While decimating science programs, the administration has modestly increased funding for human spaceflight by 10 percent, allocating $7 billion for lunar missions and $1 billion for Mars exploration, though it simultaneously plans to retire the Space Launch System rocket and Orion capsule after Artemis III and cancel the Gateway lunar station. The proposed cuts would also eliminate nearly one-third of NASA's workforce, with particularly severe impacts at the Jet Propulsion Laboratory in California and Goddard Space Flight Center in Maryland.

These cuts represent far more than budget reductions—they constitute a systematic dismantling of America's scientific enterprise across the federal government. The Centers for Disease Control, the Department of Education, the Environmental Protection Agency, the National Aeronautics and Space Administration, the U.S. Department of Agriculture, and the U.S. Government Services all face crippling reductions ranging from 24 to 55 percent, with some programs eliminated entirely. Should Congress fail to reject these unprecedented cuts, the United States will abandon its position as the global leader in scientific discovery and innovation, ushering in what future historians may recognize as the beginning of a new Dark Age—not just for America, but for a world that has long depended on U.S. scientific leadership to address humanity's greatest challenges.

A century from now, satellites—if any still function—would gaze down upon a transformed Earth, bearing witness to the consequences of abandoning scientific stewardship. Where once they tracked the careful management of forests and watersheds, they would see unchecked erosion and vanished ecosystems. Where once they warned of approaching hurricanes and helped communities prepare, they would observe the scars of disasters that struck without warning. The precise agricultural fields they once helped optimize would have given way to depleted soils and failed harvests. The ice sheets they meticulously measured would exist only in archived images, their meltwater having reshaped every coastline. These future satellites would document not the gradual warming their predecessors predicted, but the catastrophic acceleration that followed when humanity chose to blind itself, shutting down the very instruments that served as Earth's vital signs. Looking down from their decaying orbits, they would see a planet that had exchanged knowledge for ignorance, foresight for hindsight, a clement climate for catastrophic warming, and prevention for regret—a world where the lights of scientific inquiry have gone out, leaving only the darkness of what might have been.

Acknowledgments

Thanks, as always, to John Thornton, my agent, and to Jonathan Cobb for skilled editing

Notes

Prelims

1. "Konstantin Tsiolkovsky," accessed July 24, 2024, https://www.esa.int/Science_Explorat ion/Human_and_Robotic_Exploration/Exploration/Konstantin_Tsiolkovsky.

Chapter 1

1. James Lawrence Powell, *Unlocking the Moon's Secrets: From Galileo to Giant Impact* (Oxford University Press, 2023), https://books.google.com/books?id=A83MEAAAQBAJ.

Chapter 2

1. E. E. Hale, *The Brick Moon, and Other Stories*, The Works of Edward Everett Hale. [IV] (Little, Brown, 1899), https://books.google.com/books?id=q_VEAQAAMAAJ.
2. Jules Verne, *From the Earth to the Moon, Direct in Ninety-Seven Hours and Twenty Minutes: And a Trip Round It* (Project Gutenberg, 1873).
3. "Index Translationum," in *Wikimedia*, October 7, 2022, https://en.wikipedia.org/w/ index.php?title=Index_Translationum&oldid=1114534240.
4. V. S. Savchuk, N. M. Kushlakova, and I. B. Vavilova, "Nikolai Kibalchich in the History of World Rocket-Space Technics: Discussion Questions of Domestic and World Historiography," *Kosmichna Nauka i Tekhnologiya* 25 (December 1, 2019): 70–83, https://doi.org/ 10.15407/knit2019.06.070.>
5. J. T. Andrews, *Red Cosmos: K. E. Tsiolkovskii, Grandfather of Soviet Rocketry*, Centennial of Flight Series (Texas A&M University Press, 2009), https://books.google.com/ books?id=3jO5KHOLdDcC. 38.
6. "The Collection of Works by Konstantin Tsiolkovsky (English)," accessed June 27, 2023, https://www.tsiolkovsky.org/en/cosmic-philosophy-by-tsiolkovsky.
7. Adil Siddiqi, "Challenge to Apollo," NASA History Division (National Aeronautics and Space Administration, 2000).
8. Frank Winter, *Rockets into Space* (Harvard University Press, 1990), https://archive.org/ details/rocketsintospace0000wint_g0i1/page/18/mode/2up.
9. Winter, 19.
10. Eric Bergaust, *Wernher von Braun* (Stackpole Books, 1976).
11. R. H. Goddard, "A Method of Reaching Extreme Altitudes" (Smithsonian, 1920).
12. "The Greatest Newspaper Correction Ever Written (49 Years Too Late)," Gizmodo, December 31, 2013, https://gizmodo.com/the-greatest-newspaper-correction-ever-written-49-year-1491590487.
13. R. H. Goddard, "Liquid-Propellant Rocket Development," *SMITHSONIAN MISCELLANEOUS COLLECTIONS*, 1936.

14. "The Misunderstood Professor," *Air & Space Magazine, Smithsonian Magazine*, accessed July 14, 2023, https://www.smithsonianmag.com/air-space-magazine/the-misunderstood-professor-26066829.
15. Wernher von Braun, "Pioneers of a New Age," *The New York Times*, July 17, 1969, 42.
16. Winter, *Rockets into Space*, vii.

Chapter 3

1. James Lawrence Powell, *Faith in Fallacy: A Century of State-Sanctioned Science Denial* (Oxford University Press, 2024).
2. Mike Gruntman, *Blazing the Trail: The Early History of Spacecraft and Rocketry* (American Institute of Aueronautics and Astronautics, 2004), 275.
3. James C. Hagerty, "Press Release," March 5, 2019, https://web.archive.org/web/20190305033401/https://www.eisenhower.archives.gov/research/online_documents/igy/1955_7_29_Press_Release.pdf.

Chapter 4

1. Gruntman, *Blazing the Trail*, 335.
2. Paul Dickson, *Sputnik: The Shock of the Century* (Berkley Books, 2003), 137.
3. Dickson, 112.
4. You can hear Sputnik's beeps at https://soundcloud.com/nasa/sputnik-beep.
5. Dickson, *Sputnik*, 146.
6. Bergaust, *Wernher von Braun*.
7. J. B. Medaris, *Countdown for Decision* (Creative Media Partners, LLC, 2021), https://books.google.com/books?id=qEWtzgEACAAJ. 166.
8. Dickson, *Sputnik*, 174.

Chapter 5

1. William H. Guier and George C. Weiffenbach, "Genesis of Satellite Navigation," *Johns Hopkins Apl Technical Digest* 18, no. 2 (1997).
2. Guier and Weiffenbach, 180.
3. "Sci-Fi Author Arthur C. Clarke Predicted GPS, Satellite TV and Cellphones—The Tech Journal," July 27, 2010, https://thetechjournal.com/off-topic/sci-fi-author-arthur-c-clarke-predicted-gps-satellite-tv-and-cellphones.xhtml.

Chapter 7

1. J. L. Powell, *The 2084 Report: A Novel of the Great Warming* (Atria Books, 2021), https://books.google.com/books?id=L2U0EAAAQBAJ.
2. Yaojun Li et al., "Climate-Driven Acceleration of Glacier Mass Loss on Global and Regional Scales during 1961–2016," *Science China Earth Sciences* 64, no. 4 (April 2021): 589–99, https://doi.org/10.1007/s11430-020-9700-1.

3. M. Lisowski, J. C. Savage, and W. H. Prescott, "The Velocity Field along the San Andreas Fault in Central and Southern California," *Journal of Geophysical Research: Solid Earth* 96, no. B5 (May 10, 1991): 8369–89, https://doi.org/10.1029/91JB00199.

4. Kristine M. Larson, Paul Bodin, and Joan Gomberg, "Using 1-Hz GPS Data to Measure Deformations Caused by the Denali Fault Earthquake," *Science* 300, no. 5624 (May 30, 2003): 1421–24, https://doi.org/10.1126/science.1084531.

5. Diego Melgar and Gavin P. Hayes, "Characterizing Large Earthquakes before Rupture Is Complete," *Science Advances* 5, no. 5 (May 29, 2019): eaav2032, https://doi.org/10.1126/sciadv.aav2032.

6. Geoffrey Blewitt et al., "Rapid Determination of Earthquake Magnitude Using GPS for Tsunami Warning Systems," *Geophysical Research Letters* 33, no. 11 (June 2006): 2006GL026145, https://doi.org/10.1029/2006GL026145; Richard A. Kerr, "Failure to Gauge the Quake Crippled the Warning Effort," *Science* 307, no. 5707 (January 14, 2005): 201–201, https://doi.org/10.1126/science.307.5707.201.

7. Quentin Bletery and Jean-Mathieu Nocquet, "The Precursory Phase of Large Earthquakes," *Science* 381, no. 6655 (July 21, 2023): 297–301, https://doi.org/10.1126/science.adg2565.

8. "Predicting Earthquakes Hours Ahead of Time May Finally Be Possible—The Debrief," accessed April 17, 2024, https://thedebrief.org/predicting-earthquakes-hours-ahead-of-time-may-finally-be-possible.

9. Giorgio Savastano et al., "Real-Time Detection of Tsunami Ionospheric Disturbances with a Stand-Alone GNSS Receiver: A Preliminary Feasibility Demonstration," *Scientific Reports* 7, no. 1 (April 21, 2017): 46607, https://doi.org/10.1038/srep46607.

10. C. A. Neal et al., "The 2018 Rift Eruption and Summit Collapse of Kīlauea Volcano," *Science* 363, no. 6425 (January 25, 2019): 367–74, https://doi.org/10.1126/science.aav7046.

11. Kristine M. Larson, "A New Way to Detect Volcanic Plumes," *Geophysical Research Letters* 40, no. 11 (June 16, 2013): 2657–60, https://doi.org/10.1002/grl.50556.

12. Manuel Martin-Neira, "A Passive Reflectometry and Interferometry System (PARIS): Application to Ocean Altimetry," *ESA Journal* 17, no. 4 (1993): 331–55.

13. Kristine M. Larson, "GPS Interferometric Reflectometry: Applications to Surface Soil Moisture, Snow Depth, and Vegetation Water Content in the Western United States," *WIREs Water* 3, no. 6 (2016): 775–87, https://doi.org/10.1002/wat2.1167.

14. Kristine M. Larson, Richard D. Ray, and Simon D.P. Williams, "A 10-Year Comparison of Water Levels Measured with a Geodetic GPS Receiver versus a Conventional Tide Gauge," *Journal of Atmospheric and Oceanic Technology* 34, no. 2 (2017): 295–307.

Chapter 8

1. A. Blum, *The Weather Machine: A Journey Inside the Forecast* (HarperCollins, 2019), https://books.google.com/books?id=2WlwDwAAQBAJ.

2. Blum, 15.

3. J. Hill, *Weather from Above* (Smithsonian, 1991), https://books.google.com/books?id=TXW-z3qnu2wC.

4. Blum, *The Weather Machine*. 26.

5. "American Scientists Took the First Photo of Earth from Space Using Nazi Rockets," Smart News, *Smithsonian Magazine*.

6. Hill, *Weather from Above*, 4.

7. Defense Technical Information Center, *DTIC ADA307104: Inquiry into the Feasibility of Weather Reconnaissance from a Satellite Vehicle*, 1951, http://archive.org/details/DTIC_ADA307104.

8. "Harry Wexler: Father of Weather Satellites," accessed April 20, 2024, https://www.esa.int/About_Us/ESA_history/Harry_Wexler_Father_of_weather_satellites.

9. Blum, *The Weather Machine*; Harry Wexler, "Observing the Weather from a Satellite Vehicle," in *Exploring the Unknown: Selected Documents in the History of the U. S. Civil Space Program: Using Space.*, vol. III (DIANE Publishing Company, 2003), 178–83. 183.

10. Wexler, "Observing the Weather from a Satellite Vehicle," 183.

11. James Fleming, "ARTIFACTS: A 1954 Color Painting of Weather Systems as Viewed from a Future Satellite," *Bulletin of the American Meteorological Society* 88, no. 10 (October 2007): 1525–27, https://doi.org/10.1175/BAMS-88-10-1525.

Chapter 9

1. Meteorologists had long known that the weather overall is divided into systems that vary in size from about 100 km to several thousand km. These would include regions of high and low pressure with a "front" between them; tropical cyclones (hurricanes in the United States); the jet stream; and monsoons. The only way to view them from a high enough altitude to see a large system in one view is from a satellite.

2. Hill, *Weather from Above.*

3. Hill, 30.

4. "Global Views of Our Planet, Then and Now," Text.Article (NASA Earth Observatory, April 22, 2015), https://earthobservatory.nasa.gov/images/85736/global-views-of-our-planet-then-and-now#.

Chapter 10

1. Anne Braakmann-Folgmann et al., "Mapping the Extent of Giant Antarctic Icebergs with Deep Learning," *The Cryosphere* 17, no. 11 (November 9, 2023): 4675–90, https://doi.org/10.5194/tc-17-4675-2023.

2. J. Tournadre et al., "The ALTIBERG Iceberg Data Base Version 3.0, Antarctic and Arctic Data Sets," 2021, https://www.researchgate.net/profile/Jean-Tournadre/publication/351034498_The_ALTIBERG_iceberg_data_base_version_30_Antarctic_and_Arctic_data_sets/links/6080360b2fb9097c0cfdc629/The-ALTIBERGiceberg–version-30-Antarctic-and-Arctic-data-sets.pdf.data-base

3. James Lawrence Powell, *Mysteries of the Deep: How Seafloor Drilling Expeditions Revolutionized Our Understanding of Earth History* (MIT Press, 2024).

4. Craig E. Williamson et al., "Lakes and Reservoirs as Sentinels, Integrators, and Regulators of Climate Change," *Limnology and Oceanography* 54, no. 6 (2009): 2273–82.

5. Nishan Kumar Biswas et al., "Towards a Global Reservoir Assessment Tool for Predicting Hydrologic Impacts and Operating Patterns of Existing and Planned Reservoirs,"

Environmental Modelling & Software 140 (June 1, 2021): 105043, https://doi.org/10.1016/j.envsoft.2021.105043.

6. "Hydroweb – Theia," accessed December 5, 2023, https://www.theia-land.fr/en/hydroweb.

Chapter 11

1. "Study on Water Storage Change and Its Consideration in Water Balance in the Mountain Regions over Arid Northwest China—Consensus," accessed July 20, 2024, https://consensus.app/papers/study-water-storage-change-consideration-water-balance-xu/cbeffc5af50c518181c9a6470940d656/?utm_source=chatgpt.

2. Alexandra S. Richey et al., "Quantifying Renewable Groundwater Stress with GRACE," *Water Resources Research* 51, no. 7 (July 2015): 5217–38, https://doi.org/10.1002/2015WR017349.

3. http://drought.unl.edu

4. John T. Reager, Brian F. Thomas, and James S. Famiglietti, "River Basin Flood Potential Inferred Using GRACE Gravity Observations at Several Months Lead Time," *Nature Geoscience* 7, no. 8 (2014): 588–92.

Chapter 12

1. William K. Stevens, "Scientist at Work: Wallace S. Broecker; Iconoclastic Guru of the Climate Debate," *The New York Times*, March 17, 1998, sec. Science, https://www.nytimes.com/1998/03/17/science/scientist-at-work-wallace-s-broecker-iconoclastic-guru-of-the-climate-debate.html.

2. James Hansen et al., "Potential Climate Impact of Mount Pinatubo Eruption," *Geophysical Research Letters* 19, no. 2 (1992): 215–18, https://doi.org/10.1029/91GL02788.

3. Joseph C. Farman, Brian G. Gardiner, and Jonathan D. Shanklin, "Large Losses of Total Ozone in Antarctica Reveal Seasonal ClO x/NO x Interaction," *Nature* 315, no. 6016 (1985): 207–10.

4. "Statement on Signing the Montreal Protocol on Ozone-Depleting Substances," Ronald Reagan, accessed July 20, 2024, https://www.reaganlibrary.gov/archives/speech/statement-signing-montreal-protocol-ozone-depleting-substances.

Chapter 13

1. Tim DeVries, Mark Holzer, and Francois Primeau, "Recent Increase in Oceanic Carbon Uptake Driven by Weaker Upper-Ocean Overturning," *Nature* 542, no. 7640 (February 2017): 215–18, https://doi.org/10.1038/nature21068.

2. Merchant et al., "2—Global Sea Surface Temperature."

3. Jennifer K. McWhorter, Paul R. Halloran, George Roff, William J. Skirving, Chris T. Perry, and Peter J. Mumby, "The Importance of 1.5°C Warming for the Great Barrier Reef," *Global Change Biology* 28, no. 4 (2022): 1332–41, https://doi.org/10.1111/gcb.15994.

4. Terry P. Hughes et al., "Global Warming and Recurrent Mass Bleaching of Corals," *Nature* 543, no. 7645 (March 2017): 373–77, https://doi.org/10.1038/nature21707.

Chapter 14

1. "Measuring Vegetation (NDVI & EVI)," Text.Article (NASA Earth Observatory, August 30, 2000), https://earthobservatory.nasa.gov/features/MeasuringVegetation/measuring_vegetation_1.php.

2. Genesis T. Yengoh et al., "Applications of NDVI for Land Degradation Assessment," in *Use of the Normalized Difference Vegetation Index (NDVI) to Assess Land Degradation at Multiple Scales*, SpringerBriefs in Environmental Science (Springer International Publishing, 2015), 17–25, https://doi.org/10.1007/978-3-319-24112-8_3.

3. R. B. Myneni et al., "Increased Plant Growth in the Northern High Latitudes from 1981 to 1991," *Nature* 386, no. 6626 (April 1997): 698–702, https://doi.org/10.1038/386698a0.

4. Ephias M. Makaudze and Mario J. Miranda, "Catastrophic Drought Insurance Based on the Remotely Sensed Normalised Difference Vegetation Index for Smallholder Farmers in Zimbabwe," *Agrekon* 49, no. 4 (2010): 418–32.

5. Irene Winkler et al., "Assessing the Applicability of NDVI Data for the Design of Index-Based Agricultural Insurance in Bihar, India," IEEE International Geoscience and Remote Sensing Symposium (IGARSS) (IEEE, 2015), 854–57, https://ieeexplore.ieee.org/abstract/document/7325899.

6. "Mapping a Better Vintage| NASA Spinoff," accessed January 9, 2024, https://spinoff.nasa.gov/spinoff2003/er_2.html.

7. A. Cyr and S. Nebel, "Satellite and Data Logger Telemetry of Marine Vertebrates," *Nature Education Knowledge* 4, no. 2 (2013), https://www.nature.com/scitable/kno; https://www.nature.com/scitable/knowledge/library/satellite-and-data-logger-telemetry-of-marine-96643712.

8. Ramón Bonfil et al., "Transoceanic Migration, Spatial Dynamics, and Population Linkages of White Sharks," *Science* 310, no. 5745 (October 7, 2005): 100–103, https://doi.org/10.1126/science.1114898.

9. Scott R. Benson et al., "Post-Nesting Migrations of Leatherback Turtles (Dermochelys Coriacea) from Jamursba-Medi, Bird's Head Peninsula, Indonesia," *Chelonian Conservation and Biology* 6, no. 1 (May 2007): 150–54, https://doi.org/10.2744/1071-8443(2007)6[150:PMOLTD]2.0.CO;2.

10. Carl Zimmer, "7,000 Miles Nonstop, and No Pretzels," *The New York Times*, May 24, 2010, sec Science, https://www.nytimes.com/2010/05/25/science/25migrate.html.

11. Jim Robbins, "The Godwit's Long, Long Nonstop Journey," *The New York Times*, September 20, 2022, sec. Science, https://www.nytimes.com/2022/09/20/science/migratory-birds-godwits.html.

12. Clarke et al., "Cetacean Strandings from Space."

13. Song Gao, "Geospatial Artificial Intelligence (GeoAI)," Vol. 10. New York: Oxford University Press, 2021. https://doi.org/10.1093/obo/9780199874002–0228.

14. Joris Laborie et al., "Estimation of Total Population Size of Southern Elephant Seals (*Mirounga Leonina*) on Kerguelen and Crozet Archipelagos Using Very High-Resolution Satellite Imagery," *Frontiers in Marine Science* 10 (2023), https://www.frontiersin.org/articles/10.3389/fmars.2023.1149100.

15. Peter T. Fretwell et al., "An Emperor Penguin Population Estimate: The First Global, Synoptic Survey of a Species from Space," *PloS One* 7, no. 4 (2012): e33751.

16. Peter Fretwell, "Four Unreported Emperor Penguin Colonies Discovered by Satellite," *Antarctic Science*, January 20, 2024, 1–3, https://doi.org/10.1017/S0954102023000329.

17. Michael Perras and Siike Nebel, "Satellite Telemetry and Its Impact on the Study of Animal Migration," *Nature Education Knowledge* 3, no. 12: 4.

18. Roland Kays and Martin Wikelski, "The Internet of Animals: What It Is, What It Could Be," *Trends in Ecology & Evolution* 38, no. 9 (September 1, 2023): 859–69, https://doi.org/10.1016/j.tree.2023.04.007.

19. Roland Kays, Margo Crowfoot, Walter Zetz, and Martin Wikelski, "Terrestrial Animal Tracking as an Eye on Life and Planet," *Science* 348, no. 6240: 2478-1–2478-9.

20. Scott LaPoint et al., "Animal Behavior, Cost-Based Corridor Models, and Real Corridors," *Landscape Ecology* 28, no. 8 (October 2013): 1615–30, https://doi.org/10.1007/s10980-013-9910-0.

21. Terrie M. Williams et al., "Instantaneous Energetics of Puma Kills Reveal Advantage of Felid Sneak Attacks," *Science* 346, no. 6205 (October 3, 2014): 81–85, https://doi.org/10.1126/science.1254885.

22. Jeremy J. Cusack et al., "Weak Spatiotemporal Response of Prey to Predation Risk in a Freely Interacting System," *Journal of Animal Ecology* 89, no. 1 (2020): 120–31, https://doi.org/10.1111/1365-2656.12968; "Yellowstone Elk Don't Budge for Wolves Say Scientists," EurekAlert!, accessed May 6, 2024, https://www.eurekalert.org/news-releases/509699.

23. Toni K. Ruth et al., "Large-Carnivore Response to Recreational Big-Game Hunting along the Yellowstone National Park and Absaroka-Beartooth Wilderness Boundary," *Wildlife Society Bulletin* 31 (2003): 1150–61.

24. "Homepage—Animal Sensors Website," accessed May 12, 2024, https://www.icarus.mpg.de/en; Jim Robbins, "With an Internet of Animals, Scientists Aim to Track and Save Wildlife," *The New York Times*, June 9, 2020, sec. Science, https://www.nytimes.com/2020/06/09/science/space-station-wildlife.html.

25. Inhee Lee et al., "mSAIL: Milligram-Scale Multi-Modal Sensor Platform for Monarch Butterfly Migration Tracking," in *Proceedings of the 27th Annual International Conference on Mobile Computing and Networking* (ACM MobiCom '21: The 27th Annual International Conference on Mobile Computing and Networking, New Orleans, Louisiana: ACM, 2021): 517–30, https://doi.org/10.1145/3447993.3483263.

Chapter 15

1. James Acker, "Ocean-Colour Radiometry and the Public," in *Why Ocean Colour? The Societal Benefits of Ocean- Colour Technology*, n.d., 3–12.

2. B. B. Cael, K. Bisson, E. Boss, S. Dutkiewicz, & S. Henson. (2023). "Global Climate-change Trends Detected in Indicators of Ocean Ecology," *Nature*, 619(7970): 551–54.

3. Stephanie Dutkiewicz et al., "Ocean Colour Signature of Climate Change," *Nature Communications* 10, no. 1 (February 4, 2019): 578, https://doi.org/10.1038/s41467-019-08457-x.

4. Chris Mays et al., "Lethal Microbial Blooms Delayed Freshwater Ecosystem Recovery Following the End-Permian Extinction," *Nature Communications* 12 (September 17, 2021): 5511, https://doi.org/10.1038/s41467-021-25711-3.

5. "Harmful Algal Blooms: Red Tide: Home| CDC HSB," August 27, 2009, https://web.archive.org/web/20090827120347/http://www.cdc.gov/hab/redtide.

6. Trevor Platt et al., "Why Ocean Colour? The Societal Benefits of Ocean-Colour Technology," *Reports and Monographs of the International Ocean-Colour Coordinating Group (IOCCG)*, 2008, http://www.compendiumkustenzee.be/en/imis-mog?module=ref&refid=134492&printversion=1&dropIMIStitle=1.

7. Jeffrey J. Polovina et al., "Forage and Migration Habitat of Loggerhead (*Caretta Caretta*) and Olive Ridley (*Lepidochelys Olivacea*) Sea Turtles in the Central North Pacific Ocean," *Fisheries Oceanography* 13, no. 1 (January 2004): 36–51, https://doi.org/10.1046/j.1365-2419.2003.00270.x.

8. Christophe Guinet et al., "Spatial Distribution of Foraging in Female Antarctic Fur Seals Arctocephalus Gazella in Relation to Oceanographic Variables: A Scale-Dependent Approach Using Geographic Information Systems," *Marine Ecology Progress Series* 219 (2001): 251–64.

9. Daniel M. Palacios, "GalCet2K: A Line-Transect Survey for Cetaceans across an Environmental Gradient off the Galápagos Islands, 5–19 April 2000," Final Report Submitted to: Galápagos National Park Service, Charles Darwin Research Station, Capitanía de Puerto Ayora, 2000, https://www.researchgate.net/profile/Daniel-Palacios-13/publication/230630820_GalCet2K_A_line-transect_survey_for_cetaceans_across_an_environmental_gradient_off_the_Galapagos_Islands_5-19_April_2000/links/00b7d52cec52cc3c83000000/GalCet2K-A-line-transect-survey-for-cetaceans-across-an-environmental-gradient-off-the-Galapagos-Islands-5-19-April-2000.pdf.

10. Sue Moore et al., "Blue Whale Habitat Associations in the Northwest Pacific: Analysis of Remotely-Sensed Data Using a Geographic Information System," *Oceanography* 15, no. 3 (2002): 20–25.

11. William J. Broad, "Satellites Spot Oceans Aglow with Trillions of Organisms," *The New York Times*, August 27, 2021, sec. Science, https://www.nytimes.com/2021/08/27/science/ocean-bioluminescent-satellite.html.

12. C. R. Darwin, *Narrative of the Surveying Voyages of His Majesty's Ships Adventure and Beagle between the Years 1826 and 1836, Describing Their Examination of the Southern Shores of South America, and the Beagle's Circumnavigation of the Globe. Journal and Remarks. 1832–1836* (London: Henry Colburn, 1839), accessed April 17, 2024, https://darwin-online.org.uk/content/frameset?pageseq=30&itemID=F10.3&viewtype=text.

13. John Edward Huth, *The Lost Art of Finding Our Way* (Belknap Press of Harvard University Press, 2013), http://archive.org/details/lostartoffinding0000huth.

14. Broad, "Satellites Spot Oceans Aglow with Trillions of Organisms."

15. Platt et al., "Why Ocean Colour?"

Chapter 16

1. Lyman Spitzer, "Report to Project Rand: Astronomical Advantages of an Extra-Terrestrial Observatory," *Astronomy Quarterly* 7, no. 3 (January 1, 1990): 131–42, https://doi.org/10.1016/0364-9229(90)90018-V.

2. A. May, *Eyes in the Sky: Space Telescopes from Hubble to Webb* (Icon Books, 2024), https://books.google.com/books?id=T7TjEAAAQBAJ. This book is the main resource for this chapter.

3. Powell, *Mysteries of the Deep*.

4. "Cosmic Search Issue 12 Page 8—The First 50 Years of Radio Astronomy, Part 1: Karl Jansky and His Discovery of Radio Waves from Our Galaxy by John Kraus," accessed February 20, 2024, http://www.bigear.org/CSMO/HTML/CS12/cs12p08.htm.

5. May, *Eyes in the Sky.*

6. Powell, *Unlocking the Moon's Secrets.*

7. Nancy Grace Roman, "Nancy Grace Roman and the Dawn of Space Astronomy," *Annual Review of Astronomy and Astrophysics* 57, no. 1 (August 18, 2019): 1–34, https://doi.org/10.1146/annurev-astro-091918-104446.

8. "Dr. Nancy Grace Roman (1925–2018)," accessed April 3, 2024, https://science.nasa.gov/people/nancy-roman.

9. American Institute of Physics, "Nancy G. Roman," September 24, 2021, https://www.aip.org/history-programs/niels-bohr-library/oral-histories/4846.

10. May, *Eyes in the Sky.*

11. Roman, "Nancy Grace Roman and the Dawn of Space Astronomy."

12. William C. Keel, *The Road to Galaxy Formation* (Springer Science & Business Media, 2007). Strictly speaking, the inverse of time.

13. "WMAP- Cosmological Constant or Dark Energy," accessed February 26, 2024, https://map.gsfc.nasa.gov/universe/uni_accel.html.

Chapter 17

1. Ars Staff, "A History of Dark Matter," *Ars Technica*, February 3, 2017, https://arstechnica.com/science/2017/02/a-history-of-dark-matter/.

2. Marco Tavani and Leigh Brookshaw, "The Origin of Planets Orbiting Millisecond Pulsars," *Nature* 356, no. 6367 (March 1992): 320–22, https://doi.org/10.1038/356320a0.

3. Pierre-Yves Martin, "Catalogue of Exoplanets," exoplanet.eu, 1995, https://exoplanet.eu/catalog.

4. Pierre-Yves Martin, "Catalogue of Exoplanets."

5. May, *Eyes in the Sky.*

6. "Exoplanets," HubbleSite, accessed February 28, 2024, https://hubblesite.org/home/science/exoplanets.

7. "NASA—Hubble Finds First Organic Molecule on an Exoplanet—NASA Science," accessed July 20, 2024, https://science.nasa.gov/missions/hubble/nasa-hubble-finds-first-organic-molecule-on-an-exoplanet.

8. Scott S. Sheppard et al., "New Jupiter and Saturn Satellites Reveal New Moon Dynamical Families," *Research Notes of the AAS* 7, no. 5 (May 2023): 100, https://doi.org/10.3847/2515-5172/acd766.

9. Rebecca Boyle, "The Race Is on to Discover the First Moon around a Planet beyond Our Solar System," *Scientific American* (2021) https://www.scientificamerican.com/article/the-race-is-on-to-discover-the-first-moon-around-a-planet-beyond-our-solar-system/.

10. May, *Eyes in the Sky*, 39.

11. N. English, *Space Telescopes: Capturing the Rays of the Electromagnetic Spectrum, Astronomers' Universe* (Springer International Publishing, 2016), https://link.springer.com/book/10.1007/978-3-319-27814-8.

12. "Einstein and Lemaître: Two Friends, Two Cosmologies . . . Inters.Org," accessed July 24, 2024, https://inters.org/einstein-lemaitre.

13. Dennis Overbye, "The Webb Telescope's Latest Stumbling Block: Its Name," *The New York Times*, October 20, 2021, sec. Science, https://www.nytimes.com/2021/10/20/science/webb-telescope-astronomy-homophobia.html.

14. May, *Eyes in the Sky.*

15. Adam G. Riess et al., "Crowded No More: The Accuracy of the Hubble Constant Tested with High Resolution Observations of Cepheids by JWST" (arXiv, July 28, 2023), https://doi.org/10.48550/arXiv.2307.15806.

16. "Webb Unlocks Secrets of One of the Most Distant Galaxies Ever Seen," accessed April 4, 2024, https://science.nasa.gov/missions/webb/webb-unlocks-secrets-of-one-of-the-most-distant-galaxies-ever-seen.

17. "Webb Finds Evidence for Neutron Star at Heart of Young Supernova Remnant," Webb, accessed March 2, 2024, https://webbtelescope.org/contents/news-releases/2024/news-2024-112; C. Fransson et al., "Emission Lines Due to Ionizing Radiation from a Compact Object in the Remnant of Supernova 1987A," *Science* 383, no. 6685 (February 23, 2024): 898–903, https://doi.org/10.1126/science.adj5796.

18. Carl Zimmer, "Webb Telescope Will Look for Signs of Life Way Out There," *The New York Times*, July 2, 2022, sec. Science, https://www.nytimes.com/2022/07/02/science/webb-telescope-exoplanets-atmosphere.html.

19. Nicholas F. Wogan et al., "JWST Observations of K2-18b Can Be Explained by a Gas-Rich Mini-Neptune with No Habitable Surface," *The Astrophysical Journal Letters* 963, no. 1 (February 2024): L7, https://doi.org/10.3847/2041–8213/ad2616.

20. "Webb Image Release—Webb Sp Space Telescope GSFC/NASA," accessed August 22, 2024. https://webb.nasa.gov/index.html.

Chapter 19

1. "Satellite Database| Union of Concerned Scientists," accessed April 18, 2024, https://www.ucsusa.org/resources/satellite-database.

2. "Satellites Are Polluting the Stratosphere," *The Economist*, accessed March 6, 2025, https://www.economist.com/science-and-technology/2025/03/05/satellites-are-polluting-the-stratosphere.

3. Committee on Achieving Science Goals with CubeSats et al., *Achieving Science with Cube-Sats: Thinking Inside the Box* (National Academies Press, 2016), https://doi.org/10.17226/23503.

4. Erik Kulu, "Nanosats Database," Nanosats Database, accessed April 8, 2024, https://www.nanosats.eu/index.html.

5. Graeme Stephens et al., "The Emerging Technological Revolution in Earth Observations," *Bulletin of the American Meteorological Society* 101, no. 3 (March 1, 2020): E274–85, https://doi.org/10.1175/BAMS-D-19-0146.1.

6. Justin R. Mansell et al., "LightSail 2 Solar Sail Control and Orbit Evolution," *Aerospace* 10, no. 7 (2023): 579.

7. Martijn de Kok, Jose Velazco, and Mark Bentum, "CubeSat Array for Detection of RF Emissions from Exoplanets Using Inter-Satellite Optical Communicators," in *2020 IEEE Aerospace Conference*, 2020, 1–12, https://doi.org/10.1109/AERO47225.2020.9172296.

8. "Satellites Are Polluting the Stratosphere."

9. Staff, "China Confirms Anti-Satellite Missile Test," *The Guardian*, January 23, 2007, sec. Science, https://www.theguardian.com/science/2007/jan/23/spaceexploration.china.

10. Katrina Miller, "A Dead Russian Satellite Broke into More Than 100 Pieces in Space." *The New York Times*, June 27, 2024, sec. Science. https://www.nytimes.com/2024/06/27/science/russian-satellite-debris-iss.html.

11. William J. Broad, "New Satellites That Orbit the Earth at Very Low Altitudes May Result in a World Where Nothing Is Really off Limits," n.d.

Index